UNITEXT for Physics

Series editors

Michele Cini, Roma, Italy
Attilio Ferrari, Torino, Italy
Stefano Forte, Milano, Italy
Guido Montagna, Pavia, Italy
Oreste Nicrosini, Pavia, Italy
Luca Peliti, Napoli, Italy
Alberto Rotondi, Pavia, Italy
Paolo Biscari, Milano, Italy
Nicola Manini, Milano, Italy

UNITEXT for Physics series, formerly UNITEXT Collana di Fisica e Astronomia, publishes textbooks and monographs in Physics and Astronomy, mainly in English language, characterized of a didactic style and comprehensiveness. The books published in UNITEXT for Physics series are addressed to graduate and advanced graduate students, but also to scientists and researchers as important resources for their education, knowledge and teaching.

More information about this series at http://www.springer.com/series/13351

Rosa Poggiani

High Energy Astrophysical Techniques

 Springer

Rosa Poggiani
University of Pisa
Pisa
Italy

ISSN 2198-7882   ISSN 2198-7890   (electronic)
UNITEXT for Physics
ISBN 978-3-319-83122-0   ISBN 978-3-319-44729-2   (eBook)
DOI 10.1007/978-3-319-44729-2

Printed on acid-free paper

This Springer imprint is published by Springer Nature
The registered company is Springer International Publishing AG
The registered company address is: Gewerbestrasse 11, 6330 Cham, Switzerland

In loving memory of my mother Anna, who never gave up and continued looking up at the sky

Preface

The high energy part of the electromagnetic spectrum and the domain of cosmic rays, neutrinos, gravitational waves, dark matter arc the most recent additions to multimessenger astronomy, the world of astroparticle physics. Since high energy astrophysics involves both photons and particle probes, there is a natural splitting of observational techniques into different astronomies: X-ray, gamma ray, with cosmic rays, neutrinos, gravitational waves. Astroparticle physics is an observational science, that uses instrumentation whose elements have been originally conceived for physics at accelerators or designs new detectors for the search of rare events. The astroparticle instrumentation is built and tested in the laboratory by the astronomers. Thus astroparticle students must undergo a double training, in experimental physics and in observational astronomy. The practical information the students need is split into different places: detectors are described in high energy experimental textbooks, telescopes in high energy astrophysics textbooks, and the most recent advances are found in specialized literature. The book aims to present the instrumentation and the techniques of high energy observational astrophysics and the guidelines to plan, execute, and analyze the observations. This textbook is based on several years of teaching the courses of Astrophysical Techniques to graduate students at the University of Pisa who are specializing in astrophysics. The field of high energy astrophysics is rapidly evolving, as shown by the recent discovery of gravitational waves. The textbook is a snapshot of the current status of the art of observational technologies and presents their foreseen evolution. The textbook starts with radiation–matter interactions and discusses their impact on the design of detectors. Detectors are first presented, since they are the building blocks of the advanced instrumentation described later. The word telescope acquires a new meaning, compared to the optical domain: the telescope can use grazing incidence or be a collimator or be a combination of different instruments. The following chapters present the different astronomies that belong to high energy astrophysics, starting with the high energy region of electromagnetic spectrum and presenting astrophysics with cosmic rays, neutrinos, gravitational waves, and the searches for dark matter. Each chapter contains the orders of magnitude of the signals to be detected to link instrumentation to the astrophysics context.

The book is divided into different parts. The first part introduces the fundamentals of the field. Chapter 1 discusses the information carriers, the high energy photons, and the particle probes: cosmic rays, neutrinos, gravitational waves, dark matter; the observational windows and their constraints on the observatory site are discussed. Chapter 2 presents the radiation–matter interactions of charged particles and photons, the fundamentals tools to design the detectors. Chapter 3 discusses the interactions of the information carriers with the media encountered during the travel to the observer, that determine the observational horizons: surprisingly, more energetic particles are not necessarily able to travel longer distances. After the journey in space, high energy photons and particles encounter the terrestrial atmosphere, that acts as a large-volume calorimeter, allowing the detection of the particle showers they produce with ground-based arrays. The second part of the book describes the world of detectors. Chapter 4 presents the general characteristics of detectors; detectors for high energy particles often have intrinsic resolution capability. The single detectors are the building blocks of the astroparticle instrumentation discussed later. Chapter 5 discusses the properties of the detectors based on ionization in gases and liquids: ionization chambers, proportional counters, Geiger counters, multi-wire proportional chambers, drift chambers, liquid ionization detectors. Chapter 6 describes the scintillation detectors, materials that produce small amounts of light when hit by radiation; the light emitted by scintillators is collected by photomultiplier tubes. Chapter 7 presents the detectors based on ionization in solid-state materials, that provide imaging and spectroscopic capabilities within small volumes. Chapter 8 deals with Cherenkov and transition radiation detectors, used for particle identification. Chapter 9 discusses the calorimeters for measuring the energy of particles through their absorption in a material. Chapter 10 is a revisitation of the detectors described before from the point of view of the measurement of physical observables of photons and particles, in view of integrating them into complex instruments. The third part of the book describes the instrumentation for the different domains of high energy astrophysics. Chapter 11 deals with X-ray astronomy, performed with space-based observatories; the telescopes are based on grazing incidence or, at high energies, on collimators or coded aperture masks. Chapter 12 discusses gamma-ray astronomy and the different techniques to observe the low energy side (up to tens GeV) with space-based observatories and the high energy side (above some tens GeV) with ground-based arrays. Chapter 13 discusses the astronomy based on cosmic rays, a combination of space-based observatories at low energies and of ground-based arrays at high energies. The instrumentation used in gamma and cosmic ray investigations shows close similarities with the instrumentation at particle accelerators. Chapter 14 presents neutrino astronomy and the techniques to detect neutrinos with different energies. Chapter 15 presents the youngest astronomy, gravitational wave astronomy, which was born during the writing of this book. The different techniques used for the search of gravitational waves and the interferometers that achieved the first detection are discussed. Chapter 16 addresses the dark side of the Universe, the searches for dark matter, both direct and indirect, and the dark energy. Chapter 17 discusses the topic of observing in high energy astrophysics, with a discussion of the signal-to-noise ratio and the techniques of data analysis for the different astronomies discussed in the previous chapters. Chapter 18 discusses high energy astrophysics as a part of the multiwavelength and multimessenger astrophysics. The Web links to instrumentation are provided in the related chapters. Reference monographs are listed at the end of each chapter.

I am very grateful to several people. I thank Dario Grasso for his support and the discussions about astroparticle physics, Ivan Bruni for the support and the discussions about astronomical instrumentation, Scilla Degl'Innocenti for her support and the discussions about theoretical astrophysics, Andrea Macchi for the conversations about physics and book writing, Valentina Cettolo, Antonio Marinelli, Ignazio Bombaci. I am grateful to my advisor and mentor Gabriele Torelli, with whom I started my physicist career, to Franco Giovannelli for the interactions about physics and astronomy, to Rita Mariotti and Paolo Pancani for their support. I thank my colleagues in Virgo and LIGO for the years in the gravitational wave science. I thank the students who attended my courses at the Department of Physics of University of Pisa, for their interest and their questions. Many thanks to the technicians of the student laboratories. I am deeply indebted with Marina Forlizzi and Barbara Amorese at Springer, for their professional and kind support during the writing of this book, from the initial concept to the final version.

Last, but not least, I thank my mother Anna who has shared with me the dream of this book, but could not see it in the printed version. Without her lifelong support and encouragement, this book would not exist.

Pisa, Italy   Rosa Poggiani
July 2016

Contents

Part II   The World of Detectors

Part I
The Basics

Chapter 1
Setting the Scene: High Energy Photons and Particles

After discussing the peculiar nature of astronomy as an observational science, this chapter describes the properties of the high energy side of the electromagnetic spectrum and of other astroparticle physics probes, cosmic rays, neutrinos, gravitational waves, dark matter, the observational windows and the constraints on the sites of observatories. According to the type of probe and to its energy, experiments can be ground based (TeV gamma rays, Ultra High Energy Cosmic Rays, gravitational waves) or underground based (neutrinos) or operating on board of balloons or satellites (X-rays, GeV gamma rays, low energy cosmic rays, gravitational waves). The technical issues related to the site of the observatory will be presented, together with the astronomical and local backgrounds to observations. A short primer about astronomical coordinates is given for completeness.

1.1 Astronomy as an Observational Science

Astronomy is an observational science. The astrophysical events cannot be reproduced in the laboratory and measured several times to improve the statistics, they can only be observed without the possibility of altering their physical properties or controlling the observing conditions. Since high energy astronomies are the most recent additions to the world of astrophysics, the observations of science targets are more numerous on the side of classical astronomy, in the optical, infrared and radio domains, also for very classical sources, such as the Crab. The world of transients has been dominated in the past by the first detections performed with optical telescopes. To date, high energy observatories on Earth and in space are providing timely observations of new events. Optical astronomy provides a large amount of information and is the main ingredient of the follow-up of interesting events in the astroparticle domain. Today, the astronomical science has become multimessenger astrophysics, that combines the information of multiwavelength astronomy from the electromagnetic spectrum with the information of particle probes: cosmic rays, neutrinos, gravitational waves, dark matter.

© Springer International Publishing Switzerland 2017
R. Poggiani, *High Energy Astrophysical Techniques*,
UNITEXT for Physics, DOI 10.1007/978-3-319-44729-2_1

1.2 The Electromagnetic Spectrum: The High Energy Side

The electromagnetic spectrum deals with radiation spanning more than twenty order of magnitude in frequency, from the radio waves to the gamma rays [3, 6]. While the radio, infrared, optical and ultraviolet radiation is described as a function of the wavelength, the more energetic X-rays and gamma rays are described as a function of energy, usually measured in electron Volt (eV). High energy astrophysics deals with the part of the electromagnetic spectrum starting with the ultraviolet part and ending with the gamma rays, and can be approximately divided into regions marked by different observational techniques and instruments:

- Ultraviolet (UV): wavelengths in the range 10–300 nm
- X-rays: energy in the range 1 keV–1 MeV
- Gamma rays: energy above 1 MeV

The Earth atmosphere is opaque for high energy electromagnetic radiation, that is not able to reach the ground. The curve in Fig. 1.1 shows the atmospheric fraction (left) and the altitude (right) at which one half of the incident radiation is transmitted [4]. The ultraviolet radiation is blocked by the presence of the ozone layer. The X-rays are easily absorbed through the photoelectric effect, since their energy is close to the binding energy of the innermost electrons of atoms. The gamma rays undergo scattering through the Thomson and the Compton effects and disappear in the pair production process. The details of the interaction processes are discussed in Chaps. 2, 3. The high energy direct observations must be performed with space based or balloon borne observatories. Photons with an energy above a few hundreds

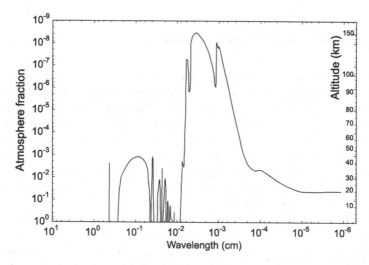

Fig. 1.1 Atmospheric absorption of photons by the Earth atmosphere as a function of their wavelength. The *curve* reports the atmospheric fraction (*left*) and the altitude (*right*) where one half of the incident radiation is transmitted; data from [4]

GeV produce cascades of particles in the atmosphere (Chap. 3), whose components can be detected by ground based instruments.

A clue to the production processes of high energy photons is given by modeling the radiation with a black body distribution, whose energy density as a function of the frequency v is:

$$B_v(T) = \frac{8\pi h v^3}{c^3} \frac{1}{e^{\frac{hv}{kT}} - 1} \qquad (1.1)$$

The effective temperatures corresponding to X-rays and gamma rays ranges from 10^6 to 10^9 K, thus the involved physical processes likely do not have a thermal origin. The emission of gamma rays does not imply that the whole source is in thermal equilibrium and acting as a black body. Gamma ray astronomy investigates the most energetic environments, from neutron stars to black holes and to the interactions of cosmic rays. Its domain spans several orders of magnitude in energy. The direct detection of gamma rays is performed with space based observatories for energies up some tens GeV. The low fluxes at higher energies preclude direct detection, that is replaced by the detection of the shower of secondary particles produced by the interaction of the gamma ray with the atmosphere. The two approaches are the subject of Chap. 12. The techniques to observe X-rays are discussed in Chap. 11.

1.3 Cosmic Rays, Neutrinos, Gravitational Waves, Dark Matter

The cosmic rays are protons and nuclei produced in high energy processes. Being charged particles, they interact with the magnetic field of the interstellar medium and when detected, they show an isotropic distribution. Only cosmic rays at very high energies undergo a small deflection and can be potentially associated to an astronomical source. Primary cosmic rays are protons, nuclei and electrons that are originally produced by an astrophysical source. Secondary cosmic rays are the products of their interactions with the interstellar medium or the Earth atmosphere. The cosmic rays with energies up to about 10^{14} eV are detected by instrumentation on board of balloons or satellites, allowing the determination of their chemical composition. The low fluxes at higher energies requires indirect detection techniques, through the detection of the particles of the cascade produced by the interaction with the atmosphere (Chap. 3). The two techniques will be discussed in detail in Chap. 13.

The neutrinos are very light neutral particles that interact only through weak interactions. The interaction cross sections are very small, thus they require detectors with very large volumes (Chap. 14). The small cross section ensures that the neutrinos preserve the information of the original source. The high energy neutrinos produced by active galactic nuclei can be detected through the detection of the Cherenkov radiation of secondary particles produced by their interaction within a large amount of transparent material, typically water or ice. The atmospheric neutrinos are an unavoidable background in neutrino telescopes. The neutrinos emitted by the Sun and in supernova explosions are at low energies and demand different techniques. The neutrino observations are described in Chap. 14.

Gravitational waves are perturbations in the spacetime that are produced by a variation of the quadrupole moment of the mass distribution of the emitter. The effects of a gravitational wave on a circular ring of free test masses is a deformation to an elliptical shape. The amplitude of the gravitational wave is given by the relative displacement of the test masses. The fundamentals of gravitational wave detection are discussed in Chap. 15.

The existence of dark matter is suggested by the rotation curves of spiral galaxies, but its nature is unknown. Direct searches of dark matter are performed at underground laboratories. Indirect searches look for the products of the annihilation of dark matter candidates. Dark energy has been suggested to explain the accelerated expansion of the Universe, but its nature is still unknown. The observational techniques related to dark matter and dark energy are described in Chap. 16.

1.4 Backgrounds

All astrophysical observations are subjected to backgrounds with different origins. The *Extragalactic Background Light* is the integral of the light emitted across the electromagnetic spectrum by early cosmological processes and by unresolved sources or diffuse regions (see [1] and references therein). The intensity of the background in the high energy region is reported in Fig. 1.2. The gamma ray background has been

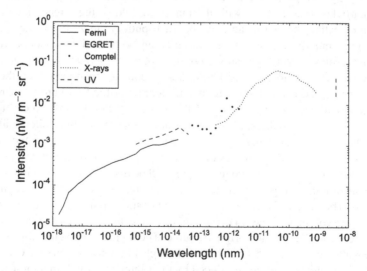

Fig. 1.2 Intensity of the extragalactic background versus the wavelength in the high energy region of the electromagnetic spectrum; data from [1]

measured by EGRET (40–10 GeV) and Fermi (100 MeV–800 GeV). The most relevant components of the background are active galactic nuclei. A possible contributor to the gamma ray background is dark matter, through the decay or the annihilation (Chap. 16). The X-ray region has been investigated by a large number of missions. The soft X-ray background is explained by the thermal emission of a gas at 10^6 K, associated to bright active galactic nuclei. The hard X-ray background is consistent with thermal bremsstrahlung radiation at 40 keV. The UV observations of the background are scarce, due to the absorption of neutral hydrogen. Other sources of background are discussed in the following chapters with reference to each kind of astronomy.

1.5 Observatory Sites

The observatories for astroparticle physics are located in a large variety of sites for the detection of a specific information carrier at a chosen energy [2, 3, 6]. The ground based observatories are used to detect very high energy particles or photons through the particles produced by their interactions with the atmosphere (Chap. 3). Differently from optical astronomy, where it produces the extinction of radiation and the degradation of images secured with telescopes, the atmosphere acts as an active medium for detection. The instrumentation to detect high energy photons and cosmic rays will be presented in Chaps. 12, 13. All direct measurements of the fluxes of high energy cosmic rays and high energy photons or of more exotic probes, such as dark matter candidates and antiparticles, must be performed in space, on board of satellites. The large cost of launch sets a strong constraint on the mass of the instrumentation, while the extreme environment of space demands instrumentation able to survive the vibrations at launch and the thermal excursions larger than one hundreds degrees in orbit. Other observatories are placed on board of balloons, an intermediate solution with a drastic drop in the cost. The search of rare events, such as the detection of neutrinos or dark matter, requires a strong reduction of environmental backgrounds and a very low environmental radioactivity. The preferred sites are underground, the heritage of the searches for the proton decay.

1.6 A Primer in Astronomical Nomenclature and Coordinates

The nomenclature of astronomical objects is very rich [3, 5]. The sources are often identified by their value in a sequence of a catalog. The catalog is labeled by an acronym of a few letters, the sequence can be a number or a short form of the astronomical coordinates. New surveys add new acronyms to the nomenclature. An example of the combination of an acronym with a sequence number is the object HD

209458, where HD labels the Henry Draper Catalog and the numeric code is the index number within the catalog. The object 1ES 2344 + 514 belongs to the catalog by the Einstein Slew Survey, while the name last part is related to its coordinates. Often the date is used to label transient events, such as the Gamma Ray Bursts, that are ordered using the standard GRByymmdd, adding the letters b, c, etc. for additional events in the same day. The discovery of gravitational waves has triggered the same kind of classification: the first observed event has been named GW150914.

The *equatorial system* is the most important coordinate system for astronomy [3, 5]. The equatorial coordinates of most objects are slowly varying, thus they can be considered as fixed stars. The equatorial system is an approximation to an inertial reference system, since it is almost fixed in space. The equivalent of the longitude and latitude are the *right ascension* and the *declination*, defined by the Greek symbols α and δ. The declination measures the altitude above the Celestial Equator, the projection of the Earth Equator on the celestial sphere; it increases in the Northern direction and decreases in the Southern direction. Right ascension starts from a reference point, the *First Point of Aries*, the equivalent of Greenwich longitude on Earth, increasing in the Eastern direction. Right ascension and declination can be measured in angular units. The declination increases from 0 to 90° in the Northern direction and decreases from 0 to −90° in the Southern direction. Historically, the right ascension is measured in time units, dividing 360° into 24 h. The right ascension increases from 0 to 24 h moving from the first point of Aries in the Eastern direction. The meridian of the observer intercepts a different right ascension value in time. The Hour Angle *HA* varies from −12 to +12 h; it is zero when the object is at meridian and changes from negative to positive values when it moves from East to West during the diurnal motion. The fundamental celestial reference system is the *International Celestial Reference System* (ICRS), a catalog of the positions of some hundreds extragalactic radio sources with an uncertainty in position of about 0.5 milli arc seconds.

The instant in time is called *epoch*, with the standard ordering year, month, day, hours, minutes, second, see 2016 January 20, 10 h 18 min 21 s. Astronomical coordinates are defined with reference to an epoch, Julian Date J2000, on 2000 January 1, 00 h 00 min 00 s. The astronomical time is based in the *Universal Time* (UT), the local time at Greenwich; local time is computed starting from Greenwich time according to the time zone of the observer. An alternative date system is used to make easier the calculation of the distance in time between two events. The *Julian Date* (JD) system has the origin at the First of January 4713 BC at noon, to have positive numbers for all dates in the history of astronomy. The Julian Day 2,450,000 occurred on 1995 October 9 at 12 h. The Julian Date system is sometimes replaced by the *Modified Julian Date* (MJD) system, where $MJD = JD - 2,400,000.5$. The *Sidereal Time* (ST) is defined as the right ascension (RA) of the observer meridian and is sum of the right ascension RA and of the hour angle HA, $ST = RA + HA$. In one year the stars cover 366 diurnal cycles, while the Sun 365 cycles only. The length of the sidereal day is 23 h, 56 min, 4 s, thus the sidereal time advances 2 h per month and has a value

Basic data :

NAME BL Lac -- BL Lac - type object

Other object types: Rad (Ref,83....), X (Ref,1AXG,...), gam (3EG,EGR....), BLL (Ref,[VV2000b],...),
                                * (PLX,UCAC2,...), Bla (Ref,[DGT2001],...), QSO (Ref,QSO,...), IR
                                (AKARI,IRAS....), V* (AN,V*....), sam (JCMTSE,JCMTSF), AGN (Ref)

ICRS coord. *(ep=J2000)* :   22 02 43.29139 +42 16 39.9803 (Radio) [0.17 0.11 0] A 2010AJ....139.1695L

FK5 coord. *(ep=J2000 eq=2000)* :   22 02 43.291 +42 16 39.98 [0.17 0.11 0]

FK4 coord. *(ep=B1950 eq=1950)* :   22 00 39.34 +42 02 08.3 [245.00 235.00 0]

Gal coord. *(ep=J2000)* :   092.5896 -10.4412 [0.17 0.11 71]

Proper motions *mas/yr* :   7.9 3.8 [4.9 4.7 0] B 2003yCat.1289....0Z

Radial velocity / Redshift / cz :   V(km/s) 19974 [~] / z(spectroscopic) 0.069 [~] / cz 20686 [~]
                                       D 2011NewA...16..503M

Parallaxes *(mas)*:   1.9 [3.5] D 1995GCTP..C......0V

Angular size *(arcmin)*:   0.0000105 0.0000057 17 (Rad) D 2010AJ....139.1713C

Fluxes (6) :   B 15.66 [~] D 2010A&A...518A..10V
                          V 14.72 [~] D 2010A&A...518A..10V
                          R 13.000 [~] E 2003yCat.2246....0C
                          J 12.201 [0.029] C 2003yCat.2246....0C
                          H 11.295 [0.037] C 2003yCat.2246....0C
                          K 10.485 [0.022] C 2003yCat.2246....0C

Fig. 1.3 SIMBAD: astronomical coordinates of BL Lac

of 0 h at midnight on 21 September. The Greenwich and local sidereal time can be obtained online using the data service at the United States Naval Observatory.[1]

There is a large variety of Internet resources for helping astronomers in the preparation of observations and in accessing public data. The *Centre de Donnés astronomiques de Strasbourg* (CDS)[2] is a repository dedicated to the management and distribution of information about astrophysical sources. The CDS comprises three main queryable databases:

- *Simbad*[3]: database for the identification of astronomical objects, that can be queried either by the name of the source or by a set of astronomical coordinates.
- *VizieR*[4]: access to more than 15000 catalogs at all wavelengths and to more than 14000 data tables from published papers; the contents can be cross correlated.
- *Aladin*[5]: an interactive sky atlas to access, visualize and perform on-line analysis of the digitized images produced by different sky surveys; different layers of data from different archives can be superimposed. Aladin is the main tool to produce the *finding charts*, the maps of the sky regions to be observed.

The result of the Simbad query for the object BL Lac and the related finding chart built with Aladin are shown in Figs. 1.3 and 1.4.

An object could be observable during the whole year, for part of the year or never, at a chosen site. The altitude *h* of an astronomical source is a function of its declination, of the hour angle and of the latitude of the observatory ϕ:

---

[1] www.aa.usno.navy.mil/data/docs/siderealtime.php.

[2] http://cds.u-strasbg.fr/.

[3] http://simbad.u-strasbg.fr/simbad/.

[4] http://vizier.u-strasbg.fr/index.gml.

[5] http://aladin.u-strasbg.fr/aladin.gml.

Fig. 1.4 Aladin: finding
chart for BL Lac

$$\sin h = \sin \delta \sin \phi + \cos \delta \cos \phi \cos HA \qquad (1.2)$$

For observations at the meridian the hour angle is zero, thus the equation becomes:

$$h = 90° - \phi + \delta \qquad (1.3)$$

The range of accessible declinations at a site is defined by $\delta \geq \phi - 90°$. The checking of the observability of an object is greatly simplified by the *Staralt* software,[6] that shows the altitude above the horizon during the whole night (Fig. 1.5).

The on line resources for astronomy are in continuing evolution. This section summarizes some starting points for astronomical investigations. Data secured with large telescopes and high energy observatories become public after 1 or 2 years and are available on line. The accessibility of archives with high quality data has the potential for new discoveries and for statistical studies of classes of objects.

The *High Energy Astrophysics Science Archive Research Center* (HEASARC)[7] is the main repository of the high energy astrophysics missions, from the extreme ultraviolet to the X-ray and gamma ray regions and includes also the experiments that have investigated the Cosmic Microwave Background (CMB) in the radio and the microwaves domain.

Observations in high energy astrophysics are accompanied by electromagnetic follow-ups in the domain of classical astronomy, the optical and infrared regions of the electromagnetic spectrum. Some related astrophysical repositories are: the

---

[6]http://catserver.ing.iac.es/staralt/.

[7]http://heasarc.gsfc.nasa.gov/.

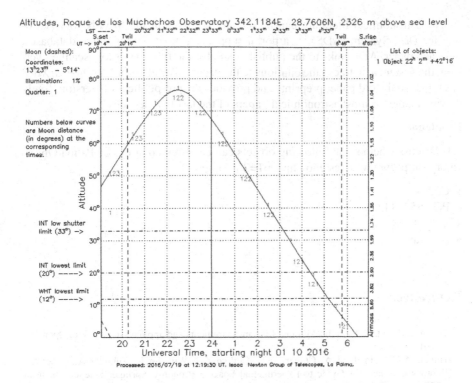

Fig. 1.5 Staralt: visibility of BL Lac on the night of 2016 October 1 at the Las Campanas Observatory

NASA/IPAC Extragalactic Database (NED)[8] with imaging and spectroscopic data of millions of objects external to the Milky Way; the *STScI Digitized Sky Survey,*[9] with the digitized data of the *Palomar Observatory Sky Surveys*; the *Sloan Digital Sky Surveys,*[10] photometric and spectroscopic surveys of millions of sources; the *Mikulski Archive for Space Telescopes (MAST),*[11] for accessing the data archives of optical, ultraviolet and near-infrared observatories, including the Hubble Space Telescope (HST) and the Digitized Sky Survey. The *Virtual Observatory* (VO)[12] is an environment managing the access to a large number of archives of ground and space based observatories. The *ASDC SED Builder Tool*[13] is an online tool to access data and display the *Spectral Energy Distribution* (SED) of astrophysical sources, combining observations over the whole electromagnetic spectrum.

---

[8]https://ned.ipac.caltech.edu/.

[9]http://stdatu.stsci.edu/cgi-bin/dss_form.

[10]http://www.sdss.org/, http://www.sdss2.org/, http://sdss3.org/.

[11]https://archive.stsci.edu/.

[12]http://www.ivoa.net, http://www.euro-vo.org/.

[13]http://www.asdc.asi.it/articles.php?id=11.

The astronomical literature can be directly accessed at the SAO/NASA Astrophysics Data System (ADS),[14] a portal supporting three bibliographic databases. For each paper, ADS links to the publisher site for the full text or to a scanned version; the system also tracks the citations to the papers. The electronic archive *arXiv*[15] has replaced the old paper preprints and provides the pre-publication version, freely downloadable, of most astronomical papers of the last two decades.

Problems

1.1 Discuss whether the following objects can be observed at the La Palma Observatory (http://www.iac.es/eno.php?op1=2\&lang=en):

- Crab
- PG 1553+113
- η Carinae
- AE Aqr

References

1. Cooray, A.: Extragalactic background light measurements and applications. R. Soc. Open Sci. 3 (2016) 150555
2. Huber, M. C. E., Pauluhn, A., Culhane, J. L., Gethyn, T. J., Wilhelm, K., Zehnder, A.: Observing Photons in Space - A Guide to Experimental Space Astronomy. Springer Science+Business Media, New York (2013)
3. Lèna, P. et al.: Observational Astrophysics. Springer-Verlag Berlin Heidelberg (2012)
4. Oda, M.: X-ray and -ray astronomy. Proc. ICRC **1**, 680 (1965)
5. Poggiani, R.: Optical, Infrared and Radio Astronomy - From Techniques to Observation. Springer (2016), doi:10.1007/978-3-319-44732-2
6. Spurio, M.: Particles and Astrophysics - A Multi-Messenger Approach. Springer International Publishing, Switzerland (2015)

---

[14]http://www.adsabs.harvard.edu/.
[15]http://arXiv.org.

Chapter 2
Radiation-Matter Interactions

The behavior of radiation and matter as a function of energy governs the degradation of astrophysical information along the path and the characteristics of the detectors. This chapter firstly presents the mechanisms of the energy loss of charged particles in matter: ionization, bremsstrahlung, Cherenkov radiation, transition radiation, nuclear reactions; in addition, the effect of multiple scattering will be summarized. Then the chapter discusses the mechanisms of photon interactions: photoelectric effect, Compton effect, pair production. The characteristic scales of electromagnetic (radiation length) and nuclear (interaction length) processes are discussed. The discussed effects are the foundation for the detectors discussed in the following. The nature of fundamental fluctuations related to the detection processes will be outlined. The interactions of radiation and particles are presented without formally deriving the relative cross sections, but rather elaborating on the critical factors leading to the detector construction.

2.1 Interactions of Charged Particles

Charged particles interact mostly with electrons and loose energy through different mechanisms [1–7]:

- Ionization and excitation of atoms encountered along the path
- Bremsstrahlung
- Cherenkov Radiation
- Transition Radiation

In addition, charged particles undergo multiple scattering that produces a series of small deviations from the path that increases its effective length.

© Springer International Publishing Switzerland 2017
R. Poggiani, *High Energy Astrophysical Techniques*,
UNITEXT for Physics, DOI 10.1007/978-3-319-44729-2_2

2.1.1   Energy Loss by Ionization

A charged particle in matter looses energy by *ionization* and *excitation* of the atoms along the path, transferring energy to the atomic electrons. The key parameter is the maximum amount of energy transferred in a single collision. The energy loss is different for heavy particles and electrons/positrons, due to their mass, that must be compared with the mass of target electrons. The energy loss per unit length of heavy charged particles, or *stopping power*, is described by the Bethe-Bloch equation [2]:

$$-\frac{dE}{dx} = 4\pi N_A r_e^2 m_e c^2 \frac{Z}{A} \frac{z^2}{\beta^2} \left[\ln \frac{2m_e c^2 \beta^2 \gamma^2}{I} - \beta^2 - \frac{\delta}{2} \right] \qquad (2.1)$$

where N_A is the Avogadro number, r_e is the classical electron radius, m_e is the electron mass, z is the charge number of the incident particle, Z, A are the atomic mass and number of the medium, I the mean excitation energy of the medium, δ the density effect correction. The mean excitation energy can be approximated by $I = 16 \cdot Z^{0.9}$ eV. Since $\frac{Z}{A}$ is close to $\frac{1}{2}$ for most media, the ionization energy loss shows a weak dependence on the material and the stopping power can be approximated by the product of the square of the particle charge and of a function of its velocity:

$$-\frac{dE}{dx} = z^2 f(\beta) \qquad (2.2)$$

The behavior of the stopping power for protons in silicon is shown in Fig. 2.1.

The Bethe-Bloch equation (Eq. 2.1) shows an energy loss proportional to $1/\beta^2$ at low energies and a logarithmic rise at high energies. The density correction takes into account the effect of the polarization of the medium that produces a screening of the electric field of the incident particle: the ionization loss approaches a plateau at high energies.

Fig. 2.1 Energy loss by ionization for protons in silicon, based on the data available at http://www.nist.gov/pml/data/star/

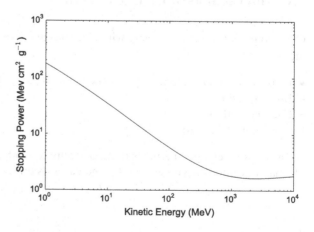

The ionization loss shows a minimum at $\beta\gamma \sim 4$. The energy loss of a *minimum ionizing particles*, called *mip*, is:

$$-\left(\frac{dE}{dx}\right)_{minimum} \sim 2\, MeV\, g^{-1}\, cm^2 \tag{2.3}$$

For example, the energy loss of a *mip* crossing 1 cm of a plastic scintillator, that has a density of about 1 g cm^{-2}, is about 2 MeV. Cosmic ray muons and relativistic particles in general are examples of the *minimum ionizing particles*. The energy loss of different particles is different in the region below the minimum of ionization and can be used to identify the type particles. The ionization loss for protons, electron and helium ions in air is shown in Fig. 2.2.

Very rarely, the energy transfer from the projectile to the electrons is large enough to allow them to produce additional ionization; the knock-on electrons are called δ-rays.

During the interaction of the charged particle with the medium, there will be fluctuations in the energy loss, whose properties depend on the thickness of the absorber material. The Bethe-Bloch equation (Eq. 2.1) describes only the average energy loss of particles. The distribution of the energy loss is a Gaussian with thick absorbers, due to the large number of collisions, but becomes asymmetrical in thin absorbers, where it is described by the *Landau distribution* [2, 6]:

$$Ł_\lambda = \frac{1}{2\pi} \exp\left[-\frac{1}{2}(\lambda + e^{-\lambda})\right] \tag{2.4}$$

where the parameter λ is given by:

$$\lambda = \frac{\Delta E - \Delta_{mp}}{\xi} \tag{2.5}$$

Fig. 2.2 Energy loss of protons, electrons and helium ions in air, based on the data available at http://physics.nist.gov/PhysRefData/Star/Text/intro.html

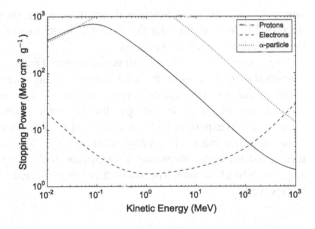

Fig. 2.3 Landau energy loss

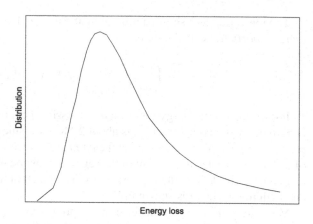

and describes the deviation of the energy loss from the most probable energy loss [2, 6]:

$$\Delta_{mp} = \xi \left[\ln \frac{2m_e c^2 \beta^2 \gamma^2}{I} + \ln \frac{\xi}{I} + 0.200 - \beta^2 - \delta \right] \qquad (2.6)$$

where $\xi = 2\pi N_A r_e^2 m_e c^2 \frac{Z}{A} z^2 \frac{d}{\beta^2}$, d is the absorber thickness in units of g cm^{-2}. The Landau distribution has a tail on the side of high energies (Fig. 2.3).

The ionization energy loss of electrons is different from the ionization of heavy particles since the projectile and the target share the same mass and differs also from the energy loss of positrons. The maximum energy transfer from incident electrons to atomic electrons is half the kinetic energy. The stopping power for electron-electron and electron-positron scattering is given by [2]:

$$-\frac{dE}{dx} = 4\pi N_A r_e^2 m_e c^2 \frac{Z}{A} \frac{1}{\beta^2} \left[\ln \frac{m_e c^2 \beta \gamma \sqrt{\gamma - 1}}{\sqrt{2}I} + F^* \right] \qquad (2.7)$$

where F^* is a different function depending on the incident particle, electron or positron [2]. We will see later that the total energy loss of electrons must include the bremsstrahlung, a radiative process.

The energy loss is the key stone of several families of detectors that will be described in the next chapters. In addition to loosing energy by ionization, a charged particle traversing a medium undergoes multiple scattering events due to the Coulomb interaction with nuclei, according to the Rutherford scattering law. During its path, the charge will experience a large number of scatterings with small deviations from the original trajectory. The *multiple Coulomb scattering* is described by the Moliere theory. Assuming that the scattering angles are small, the distribution of the scattering angles will be a Gaussian. The rms width of the projected distribution of the scattering angles is [2]:

$$\theta_{scattering} = \frac{13.6 MeV}{\beta c p} z \sqrt{\frac{d}{X_0}} \left[1 + 0.038 \ln \frac{d}{X_0} \right] \qquad (2.8)$$

where d is the medium thickness, p is the momentum of the particle in MeV/c, z its charge, X_0 is the radiation length that will be discussed in detail in the next section.

2.1.2 Bremsstrahlung

A charged particle in a medium will loose energy not only by ionization, by also by interaction with the Coulomb field of the nuclei of the material. When decelerated in the nuclear field, the particle will loose energy by emitting photons in the *bremsstrahlung* process. The energy loss by bremsstrahlung for a charge with mass m, charge number z and energy E is [2]:

$$-\frac{dE}{dx} = 4\alpha N_A \frac{Z^2}{A} z^2 \left(\frac{1}{4\pi\epsilon_0} \frac{e^2}{mc^2} \right)^2 E \ln \frac{183}{Z^{\frac{1}{3}}} \qquad (2.9)$$

where Z, A are the atomic number and mass of the medium. The dependence on the reciprocal of the squared mass of the projectile suggests that bremsstrahlung is more relevant for light particles, such as electrons. The photons are emitted within a typical angle of the order of $\sim \frac{m_e c^2}{E}$. The energy loss by bremsstrahlung is proportional to the particle energy, while the energy loss by ionization is proportional to the logarithm of energy: the bremsstrahlung will be the dominant source of losses at high energies (Fig. 2.4).

The radiation loss by bremsstrahlung is characterized by the *radiation length* X_0, the mean distance required to reduce the particle energy to a fraction $1/e$ of the initial value:

Fig. 2.4 Ionization energy loss and bremsstrahlung loss for electrons in silicon; data from http://physics.nist.gov/PhysRefData/Star/Text/ESTAR.html

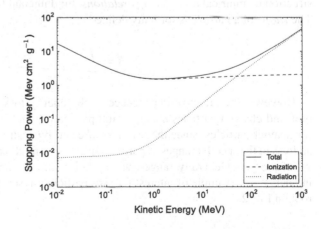

Table 2.1 Radiation length, critical energy, Moliere radius (Chap. 3), interaction length, density for materials of interest in detection (*Source* http://pdg.lbl.gov/2015/AtomicNuclearProperties/)

Material	X_0 (g cm^{-2})	E_c (MeV)	R_M (g cm^{-2})	λ_I (g cm^{-2})	ρ (g cm^{-3})
Air (1 atm)	36.62	87.92	8.83	90.1	$1.20 \cdot 10^{-3}$
Water	36.08	78.33	9.77	83.3	1.00
NaI	9.49	13.37	15.05	154.6	3.67
Polystyrene	43.79	93.11	9.97	81.7	1.06
Si	21.82	40.19	11.51	108.4	2.33
Pb	6.37	7.43	18.18	199.6	11.4
Emulsion	11.33	17.43	13.79	135.1	3.82
Liquid argon	19.55	32.84	12.62	119.7	1.40

$$-\left(\frac{dE}{dx}\right)_{radiative} = \frac{E}{X_0} \tag{2.10}$$

The radiation length X_0 depends on the material properties, the atomic number and mass:

$$X_0 = \frac{A}{4\alpha N_A Z(Z+1)r_e^2 \ln(183Z^{-\frac{1}{3}})} \tag{2.11}$$

Some typical values of the radiation length are reported in Table 2.1.

The *critical energy* is the energy where the loss rates by ionization and by bremsstrahlung are equal. An approximation of the critical energy in solids is [2]:

$$E_c \sim \frac{610}{Z+1.24} MeV \tag{2.12}$$

The critical energy is of the order of tens MeV in common materials (Table 2.1).

The complicate path of a charged particle in an absorber has triggered the construction if empirical *range-energy relations*, fundamental for the building detectors. The *range* of a particle is formally defined as:

$$R = \int_E^0 \frac{dE}{\frac{dE}{dx}} \tag{2.13}$$

However, the effective dependence of the energy loss on energy is very complex and charged particles undergo multiple scattering. The range is better defined for heavier particles, since they are less affected by scattering. Several works have addressed the specific ranges of particles in selected materials and for different interval of energies. Generally, range-energy relations can be summarized by a power law in energy. An example of interest for astroparticle physics is the range of muons in rock in Fig. 2.5.

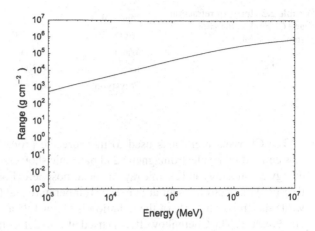

Fig. 2.5 Range of muons in rock as a function of energy; data from http://pdg.lbl.gov/2015/AtomicNuclearProperties/

2.1.3 Cherenkov Radiation

The Cherenkov radiation is emitted when a charged particle travels in a medium at a speed larger than the light speed in the medium. The velocity of light in a medium with index of refraction n is $\frac{c}{n}$. The Cherenkov effect is similar to the mechanism of the supersonic boom of a plane. The electric field of the traveling charge polarizes the medium that reverts to the previous unpolarized state afterwards, producing an electromagnetic perturbation. The combination of the perturbations events builds a single wavefront traveling along the charged particle direction at the speed of light in the medium. The angle θ_c that defines the direction of photon emission is:

$$\cos \theta_c = \frac{1}{n\beta} \tag{2.14}$$

where $\beta = \frac{v}{c}$ is the particle speed. The intensity of Cherenkov radiation per energy interval and path length interval is [3, 6]:

$$\frac{d^2N}{dEdx} = \frac{z^2\alpha}{\hbar c}\left(1 - \frac{1}{n^2\beta^2}\right) \tag{2.15}$$

where z is the particle charge. The photon yield per unit wavelength is:

$$\frac{dN}{d\lambda} = \frac{2\pi z^2\alpha}{\lambda^2}\left(1 - \frac{1}{n^2\beta^2}\right) \tag{2.16}$$

The yield of visible light is about $500\, z^2 \sin^2 \theta_c$ photons per cm. The spectrum has a peak in the high frequency region, thus the Cherenkov radiation appears as blue light. The properties of some materials used in Cherenkov detectors is presented in Table 2.2.

Table 2.2 Index of refraction and Cherenkov threshold of some materials [2]

Material	n − 1	Threshold β
Air (STP)	$2.9 \cdot 10^{-4}$	0.9997
Water	0.333	0.7501
Ice	0.309	0.7639
Polystyrene	0.59	0.6289

The Cherenkov effect is used to measure the properties of the charged particles emitted in the electromagnetic and hadronic showers (Chap. 3) induced by high energy gamma rays and cosmic rays in the atmosphere. For reference, the Cherenkov threshold of electrons in air at Standard Temperature and Pressure (STP) is 21 MeV, while the emission angle of the radiation is about 1.3° at the Standard Temperature and Pressure. The Cherenkov effect is used also to detect the charged products of the interactions of neutrinos with matter (Chap. 14).

The energy loss by Cherenkov radiation is much smaller than the loss by ionization. However, its detection is the signature that the incident particle velocity is above the threshold. The combination of the velocity with the energy information estimated by other methods provides the mass of the particle and its identification.

2.1.4 Transition Radiation

The Transition Radiation is an effect related to the polarization of the medium produced by the passage of a charged particle. When a charge in relativistic motion crosses the boundary between two media with different dielectric properties, photon emission occurs. The effect depends on the plasma frequency ω_p of the medium [2]:

$$\hbar\omega_p = \frac{m_e c^2}{\alpha}\sqrt{4\pi n_e r_e^3} \tag{2.17}$$

where n_e is the electron number density. For most materials the plasma energy is of the order of some tens eV. The spectrum of the emitted photons shows a steep decrease for energies larger than $\gamma\,\hbar\omega_p$. An estimation of the magnitude of the photon energy is about $\frac{\gamma\hbar\omega_p}{4}$. The total energy emitted when the charge crosses the boundary between the vacuum and a medium is [2]:

$$E = \frac{1}{3}\alpha z^2 \gamma\,\hbar\omega_p \tag{2.18}$$

The most part of the radiation is emitted within a cone with a semi-aperture angle of about $\frac{1}{\gamma}$. The amount of emitted energy is proportional to the relativistic factor, thus the Transition Radiation can be used for the identification of the particle type. For very fast particles, with γ of the order of thousands, the emitted photons are in the soft X-rays region of the spectrum. The number of photons emitted above a threshold $E_{th} = \hbar\omega_0$ is:

$$N_{trd} = \alpha \frac{z^2}{\pi} \left[\left(\ln \frac{\gamma \hbar \omega_p}{\hbar \omega_0} - 1 \right)^2 + \frac{\pi^2}{12} \right] \qquad (2.19)$$

2.2 Interactions of Photons

Photons are detected by the production of charged particles in their interactions with matter. The interactions of photons occur through three main processes: the *photoelectric effect*, *Compton scattering* and *pair production*. Typically, the interaction produces electrons that are later measured by their ionization loss.

2.2.1 Photoelectric Effect

The photoelectric effect is described by the process [2, 5, 6]:

$$\gamma + atom \rightarrow atom^* + e^- \qquad (2.20)$$

The atomic electrons absorb the energy of an incident photon, a process forbidden to free electrons by the momentum conservation; the process is assisted by a third body, the atomic nucleus, that deals with the recoil momentum. The cross section of photoelectric effect shows some peculiar edges superposed to a smoothly curve decreasing with energy. The edges correspond to the absorption of photons to a specific shell; the K shell gives the dominant contribution. In the regions far from the absorption edges the total photoelectric cross section is given by:

$$\sigma_{photoelectric} = \sqrt{\frac{32}{(\frac{E}{m_e c^2})^7}} \alpha^4 Z^5 \sigma_{Thomson} \qquad (2.21)$$

where E is the photon energy and $\sigma_{Thomson} = \frac{8}{3}\pi r_e^2$ is the *Thomson cross section* describing the elastic scattering of photons on electrons. The photoelectric effect is particularly important for X-ray detectors. When photoelectric effect involves an inner atomic shell, two additional effects can occur. If the empty place is filled by an electron coming from an higher shell, there will be emission of X-rays. The energy budget, if larger than the typical energy of a shell, can produce the ejection of a second electron, in the Auger effect.

2.2.2   Compton Scattering

The Compton effect is the scattering of photons on atomic electrons, that can be assumed to be almost free [2, 5, 6]:

$$\gamma + e^- \rightarrow \gamma + e^- \tag{2.22}$$

The total cross section for the Compton scattering per electron is described by the Klein-Nishina formula:

$$\sigma_{Compton}^e = 2\pi r_e^2 \left[\left(\frac{1+\varepsilon}{\varepsilon^2} \right) \left(\frac{2(1+\varepsilon)}{1+2\varepsilon} - \frac{1}{\varepsilon} \ln(1+2\varepsilon) \right) + \frac{1}{2\varepsilon} \ln(1+2\varepsilon) - \frac{1+3\varepsilon}{(1+2\varepsilon)^2} \right] \tag{2.23}$$

where $\varepsilon = \frac{E}{m_e c^2}$, E is the initial energy of the photon. For an atom with Z electrons, the scattering cross section will be $Z\sigma_{Compton}^e$. While the photoelectric effect is a pure absorption of the energy of a photon, Compton effect involves only a partial transfer of energy to the electron. The total Compton cross section can be split between the scattering cross section and the absorption cross section:

$$\sigma_{Compton}^{scattering} = \frac{E_f}{E} \sigma_{Compton}^e \tag{2.24}$$

$$\sigma_{Compton}^{absorption} = \sigma_{Compton}^e - \sigma_{Compton}^{scattering} \tag{2.25}$$

where E_f is the final energy of the photon. The relevant quantities from the point of view of detection are the differential distributions of the scattered electrons, in solid angle and in energy:

$$\frac{d\sigma_{Compton}^e}{d\Omega} = \frac{r_e^2}{2} \left(\frac{E_f}{E} \right)^2 \left[\frac{E}{E_f} - \frac{E_f}{E} - \sin^2 \theta \right] \tag{2.26}$$

$$\frac{d\sigma_{Compton}^e}{dT_e} = \frac{d\sigma_{Compton}^e}{d\Omega} \frac{2\pi}{\varepsilon^2 m_e c^2} \left[\frac{(1+\varepsilon)^2 - \varepsilon^2 \cos^2 \theta_e}{(1+\varepsilon)^2 - \varepsilon(2+\varepsilon) \cos^2 \theta_e} \right]^2 \tag{2.27}$$

where E, E_f are the energies of incident and scattered photons, θ, θ_e are the scattering angles of photon and electron, T_e the kinetic energy of electron. The energy distribution of the Compton recoil electrons is shown in Fig. 2.6.

The distribution has a cut-off, the *Compton edge*. The Compton effect is used to build detection systems for γ-rays in the regions of tens MeV. The *inverse Compton scattering*, where high energy electrons scatter on low energy photons and increase their frequency is very common in high energy astrophysics.

Fig. 2.6 Compton effect: energy distribution of the recoil electron

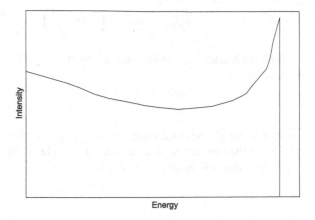

2.2.3 Pair Production

Pair production occurs when a photon converts to an electron-positron pair in the Coulomb field of a nucleus [2]:

$$\gamma + nucleus \rightarrow e^- + e^+ + nucleus \tag{2.28}$$

The process has a threshold given by the sum of the rest masses of electrons and positrons and the recoil energy of the nucleus:

$$E_{threshold} = 2m_ec^2 + 2\frac{m_e^2c^2}{m_{nucleus}} \sim 2m_ec^2 \tag{2.29}$$

Due to the large nuclei masses, the threshold can be approximated by twice the electron mass, $2m_ec^2$.

The pair production cross section depends on the degree of screening of the nuclear charge by the atomic electrons, that depends on the photon energy. The condition of no screening or complete screening depends on the value of ε, the photon energy normalized to the electron rest mass, with respect to the quantity $\frac{1}{\alpha Z^{\frac{1}{3}}}$. The pair production cross sections for no screening and complete screening are given by:

$$\sigma_{pair}^{ns} = 4\alpha r_e^2 Z^2 \left(\frac{7}{9}\ln 2\varepsilon - \frac{109}{54}\right) \tag{2.30}$$

$$\sigma_{pair}^{cs} = 4\alpha r_e^2 Z^2 \left(\frac{7}{9}\ln \frac{183}{Z^{\frac{1}{3}}} - \frac{1}{54}\right) \tag{2.31}$$

The complete screening cross section corresponds to energetic photons and does not depend on their initial energy. Dropping the last term inside the brackets of the complete screening equation, the cross section becomes:

$$\sigma_{pair}^{cs} \sim 4\alpha r_e^2 Z^2 \left(\frac{7}{9} \ln \frac{183}{Z^{\frac{1}{3}}} \right) \sim \frac{7}{9} \frac{A}{N_A} \frac{1}{X_0} \tag{2.32}$$

The length scale λ_{pair} of the pair production is related to the radiation length X_0 by the relation:

$$\lambda_{pair} = \frac{9}{7} X_0 \tag{2.33}$$

i. e. by a factor $\frac{7}{9}$. The close values of the typical lengths of the two processes will be relevant in the production of a cascade of particles started by a high energy photon or electron, that will be discussed in Chap. 3.

2.2.4   Total Cross Section for Photons

In view of designing detectors, it is useful to introduce the concept of *absorption coefficient* μ, that governs the attenuation of a photon beam in matter:

$$I = I_0 e^{-\mu x} \tag{2.34}$$

The absorption coefficient summarizes the role of the cross sections σ_i of the different interaction mechanisms σ_i:

$$\mu = \frac{N_A}{A} \sum_i \sigma_i \tag{2.35}$$

The absorption coefficient inherits the energy dependence of the physical processes mentioned above. The contribution of the photoelectric effect, Compton effect and pair production to the absorption coefficients of photons in silicon is shown in Fig. 2.7.

The photoelectric effect dominates at low energies (below some hundreds keV), the Compton scattering in the intermediate (MeV) region, while pair production is the most relevant mechanism at high energy (above some MeV). The energy dependence has a strong impact on the building of instrumentation for X-ray and gamma-ray astrophysics, as it will be discussed in Chaps. 11, 12.

In view of designing detectors for high energy photons, we briefly recall the dependence of the cross sections on the atomic number Z of the absorber material. The photoelectric effect depends on Z^5, the Compton effect on Z, the pair production on Z^2: thus detectors for high energy photons should be built with high Z materials.

For reference, we report the absorption coefficient of photons in two common detector materials, the heavy scintillator CsI and a plastic scintillator, in Fig. 2.8.

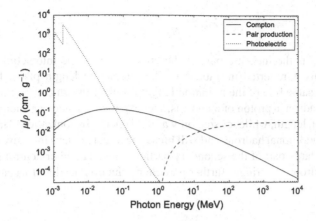

Fig. 2.7 Contribution of the photoelectric effect, the Compton effect and the pair production to the absorption coefficient of photons in silicon (data from NIST)

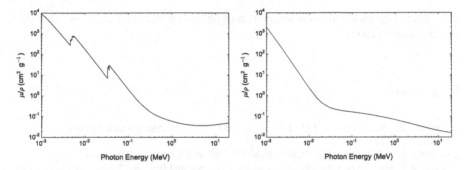

Fig. 2.8 Photon absorption coefficient for the heavy scintillator CsI (*left*) and for plastic scintillator (*right*); data from http://physics.nist.gov/PhysRefData/XrayMassCoef/tab4.html

2.3 Interaction of Hadrons

Hadrons interact with matter through strong interactions, in addition to the electromagnetic ones [2]. The majority of process belonging to this class involve inelastic scattering events that produce additional hadronic particles. The details of the interaction depend on the nature of the incident particle, a proton or a nucleus, and on its energy. The geometrical cross section for the proton-proton interaction is of the order of the squared proton radius, $\sim 10^{-26}$ cm^2; since a nucleus with atomic mass A has a size $A^{\frac{1}{3}}$, the corresponding cross section will be $A^{\frac{2}{3}}$ times larger than the proton cross section.

From the practical point of view, the large variety of strong processes accounted for by the cross section σ_h is summarized by the *hadronic interaction length* λ_{int}:

$$\lambda_{int} = \frac{A}{N_A \rho \sigma_{inelastic}}$$ (2.36)

where $\sigma_{inelastic}$ is the inelastic part of the cross section. The interaction length of some materials is reported in Table 2.1. The interaction length plays, for hadron reactions, the same role of the radiation length for electromagnetic interactions. An high energy hadron, a proton or a nucleus, will experience a nuclear interaction after one interaction length, while loosing only a small energy by ionization. The collision will produce additional hadrons and will break the target nucleus into several nuclear fragments. A large part of the secondary particles are charged or neutral pions. The secondary hadrons can trigger further particle production, leading to a cascade.

Problems

2.1 Discuss the mechanisms of energy loss for charged particles and their behavior as a function of energy.

2.2 Discuss the mechanisms of interaction of photons with matter and their behavior as a function of energy.

References

1. De Angelis, A., Pimenta, M.J.M.: Introduction to Particle and Astroparticle Physics. Springer-Verlag Italia, (2015)
2. Grupen, C. and Swartz, B.: Particle Detectors. Cambridge University Press (2008)
3. Grupen, C. and Buvat, I: Handbook of Particle Detection and Imaging. Springer-Verlag Berlin Heidelberg (2012)
4. Lèna, P. et al.: Observational Astrophysics. Springer-Verlag Berlin Heidelberg (2012)
5. Leo, W. R.: Techniques for Nuclear and Particle Physics Experiments - A How-to Approach. Springer-Verlag (1994)
6. Olive, K.A. et al. (Particle Data Group): Chin. Phys. C **38**, 090001 (2014)
7. Spurio, M.: Particles and Astrophysics - A Multi-Messenger Approach. Springer International Publishing, Switzerland (2015)

Chapter 3
Interactions of Photons and Particles Along the Path

This chapter firstly discusses the interactions of high energy photons and particles during the journey to the observer that determine the horizons for observations, the maximum distances at which high energy photons and particles can be detected. After the travel in space, photons and particles encounter the terrestrial atmosphere that acts as a large volume calorimeter and allows their detection with ground based arrays. The formation of electromagnetic showers (initiated by photons) and hadronic showers (initiated by cosmic rays) in the atmosphere is presented. The interaction of an high energy gamma ray with the atmosphere produces an electron-positron pair, that produces photons by bremsstrahlung and so on. The electromagnetic cascade of particles contains electrons, positrons and photons. The interaction of a cosmic ray (proton or nucleus) with atmosphere produces pions, kaons, etc., that decay into other particles. The hadronic shower contains electron, positrons, photons and pions, muons, neutrinos and so on.

3.1 Observational Horizons

Gamma rays and cosmic rays interact with matter along the path to the observer. Naively, the higher energy particles are expected to travel along longer distances than low energy ones. However, there are several absorption and scattering processes that affect the propagation of the probes and determine the *horizon*, the largest distance that can be accessed by observations. The probes with the highest energy are not necessarily the most penetrating.

The gamma rays interact with matter through the photoelectric effect, the Compton scattering, the pair production. The interstellar matter contribution is of the order of 10^{22}–10^{23} hydrogen atoms per cm^2, equivalent to 10^{-2} to 10^{-1} g cm^{-2}, thus it is not relevant [9]. The gamma rays also interact with photons in the environment from the Cosmic Microwave Background (CMB) at 2.7 K, the starlight radiation and so on. The photon-photon interaction produces electron-positron pairs for energies above a threshold that is inversely proportional to the energy of the gamma ray [9]. The

© Springer International Publishing Switzerland 2017
R. Poggiani, *High Energy Astrophysical Techniques*,
UNITEXT for Physics, DOI 10.1007/978-3-319-44729-2_3

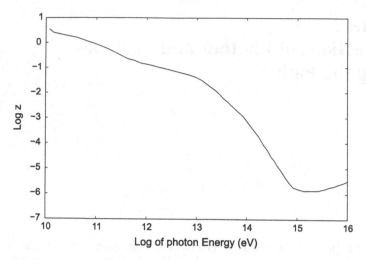

Fig. 3.1 Horizon for high energy gamma rays; data from [9]

process becomes increasingly more important for gamma rays of increasing energy; the scattering on starlight is relevant for energies below 100 GeV, where the interaction with CMB photons dominates above [9]. The horizon of high energy gamma rays is reported in Fig. 3.1. The electron-positron pairs produced in the photon-photon collisions undergo inverse Compton scattering on the environmental photons and degrade the energy of the initial gamma ray converting them to less energetic photons and producing a continuum spectrum behaving as E^{-2}.

The propagation of the cosmic rays is also affected by the Cosmic Microwave Background photons. Cosmic rays produced at cosmological distances are predicted to show a drop in the flux above the *Greisen-Zatsepin-Kuzmin* (GZK) cutoff [5], about 5×10^{19} GeV for protons. Protons interact with the CMB if their energy is larger than the threshold for the resonant production of the Δ^+ resonance at 1.232 MeV, that undergoes a prompt decay:

$$p + \gamma_{CMB} \rightarrow \Delta^+ \rightarrow \pi^+ + n \tag{3.1}$$

$$p + \gamma_{CMB} \rightarrow \Delta^+ \rightarrow \pi^0 + p \tag{3.2}$$

The threshold energy is 1.2×10^{20} eV, while the cross section of the two processes is of the order of 250 μb. The typical length for the energy loss is 30 Mpc. The limit for a cosmic ray with mass A is raised by a factor A. The GZK effect limits the horizon at which the cosmic rays can be observed to a distance smaller than 100 Mpc, depending on the mass of the primary particle. Another mechanism is pair production in the interaction with the CMB photons:

$$p + \gamma_{CMB} \rightarrow p + e^- + e^+ \tag{3.3}$$

The process has a threshold energy of about 2×10^{18} eV and a cross section of the order of 100 μb; the length scale is of the order of a few Gpc. Cosmic rays with an energy above the GZK cut-off have been observed (Chap. 13).

3.2 The Atmosphere

The atmosphere is the final barrier for incident photons and cosmic rays [9]. The basic data of the atmosphere are presented in Table 2.1 in Chap. 2 and will be recalled here. The density is about $1.20 \cdot 10^{-3}$ g cm^{-3}, while the radiation length and the interaction lengths are 36.62 and 90.1 g cm^{-2}, respectively; the critical energy is 87.92 MeV. Assuming an approximate thickness of 10 km, the integrated mass density of the atmosphere is about 1030 g cm^{-2}. The absorption coefficient of low energy gamma rays is reported in Fig. 3.2: they are quickly absorbed within a few g cm^{-2} and do not reach the ground. The low energy gamma rays must be observed with space or balloon borne observatories (Chap. 12).

The atmospheric thickness corresponds to about 28 radiation lengths and 11 interaction lengths. The primary gamma rays produce *electromagnetic showers* through the combined effect of the pair production and the bremsstrahlung, while primary cosmic rays produce *hadronic showers*. In the present chapter we will discuss the interaction of high energy photons and cosmic rays with the atmosphere. The application of the development of the electromagnetic and hadronic showers to the design of calorimeters will be presented in Chap. 9.

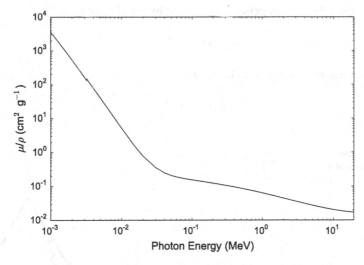

Fig. 3.2 Absorption coefficient of low energy photons in air (data from NIST)

3.3   Interaction of High Energy Photons: Electromagnetic Showers

Primary gamma rays are of special interest in astrophysics, since they are not deflected and thus they preserve the information of the position of the emitter. An high energy photon interacts with the atmosphere through the process of pair production, that produce high energy electrons and positrons that undergo bremsstrahlung. The two processes have been discussed in Chap. 2, where it has been shown that their typical scale lengths, the radiation length X_0 and the pair production length λ_{pair}, are almost identical. The incident photon will produce an electron-positron pair in the first radiation length. The two charged products will emit gamma rays by bremsstrahlung in about one radiation length. The process will continue, with high energy gamma rays producing electron-positron pairs, that in turn emit other high energy photons. The number of particles in the cascade, that is called *electromagnetic shower* [4, 10], increases exponentially. The average energy of produced particles decreases at each step of radiation length. The shower achieves a maximum in the number of component particles, that will later drop, until the absorption of all of them. The structure of an electromagnetic shower is shown in Fig. 3.3. The electromagnetic cascade has been modeled by [2, 3, 6, 8].

The electromagnetic shower contains only photons, electrons and positrons, that are strongly collimated along the direction of the primary particle, since the typical angles of the bremsstrahlung and pair production processes are of the order of $\frac{m_e c^2}{E}$. The evolution of the shower can be described as a function of the atmospheric depth X from the starting point normalized to the radiation length, i.e. of the parameter $t = \frac{X}{X_0}$. The total number of shower particles (photons, electrons, positrons) at a distance t from the shower generation is given by the Heitler model [6]:

$$N(t) = 2^t \tag{3.4}$$

with an average energy per particle equal to:

$$E(t) = E_0 2^{-t} \tag{3.5}$$

Fig. 3.3 Schematic representation of an electromagnetic shower

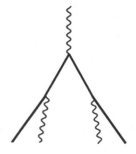

where E_0 is the energy of the primary photon. The generation of particles continues until the energy of particle drops below the critical energy E_c of the material, that sets the depth of the maximum of the shower:

$$t_{max} = \frac{\ln\left(\frac{E_0}{E_c}\right)}{\ln 2} \tag{3.6}$$

where the total number of particles is:

$$N_{max} = \frac{E_0}{E_c} \tag{3.7}$$

The number of particles at the shower maximum is a linear function of the initial energy, while the position of the maximum depends only logarithmically on energy: calorimeters are linear detectors and are compact detectors. As an example, photons with energies of 1 and 10 TeV produce 10^4 and 10^5 particles at maximum, respectively. After the maximum electrons and positrons are quickly absorbed within one radiation length, while photons are absorbed within some radiation lengths.

The depth of maximum increases logarithmically with the energy of the primary photon:

$$X^\gamma_{max} = X_s + X_0 \ln \frac{E_0}{E_c} \tag{3.8}$$

where X_s is the depth where the shower has started. The *elongation rate* relates the depth of the maximum to the initial energy:

$$D_{10} = \frac{d X^\gamma_{max}}{d(\log E_0)} \tag{3.9}$$

A popular parametrization of the number of charged particles in an electromagnetic shower has been proposed by [3]:

$$N^\gamma_e(X) = \frac{0.31}{\sqrt{\ln\frac{E_0}{E_c}}} exp\left[\left(1 - \frac{3}{2}\ln s\right)\frac{X}{X_0}\right] \tag{3.10}$$

where s is the age parameter, whose value is 1 at the maximum:

$$s = \frac{3X}{X + 2X^\gamma_{max}} \tag{3.11}$$

The size of the shower is reported in Fig. 3.4 for different energies of the primary gamma ray.

At the shower maximum, the number of charged particles and photons is given by [10]:

$$N^\gamma_{e,max} = \frac{0.31}{\sqrt{ln\frac{E_0}{E_c}}}\left(\frac{E_0}{E_c}\right) = \frac{1}{g}\left(\frac{E_0}{E_c}\right) \tag{3.12}$$

The number of charged particles in an electromagnetic shower is approximately proportional to the energy of the primary gamma ray, since the first factor is weakly dependent on the energy. The ground based arrays measure the showers by measuring their charged component. Together with the energy E_0 of the primary, the shower age summarizes the longitudinal development of the shower. As mentioned above, the emission angles of the pair production and of bremsstrahlung are very small. The *lateral development* of an electromagnetic shower is governed by the multiple scattering of electrons [10]. The width of the shower increases with increasing depth, with the most part of energy deposition occurring inside a narrow core. About 95 % of the energy of the shower is contained within a cylinder with a radius of two Moliere radii, where the *Moliere radius* is:

$$R_M = \frac{E_s}{E_c}X_0 \tag{3.13}$$

where:

$$E_s = m_e c^2 \sqrt{\frac{4\pi}{\alpha}} \tag{3.14}$$

The Moliere radius of air is about 8.8 g cm^{-2} at 1 atm, i.e. in the range 70–80 m, and is larger at higher altitudes. Since the Moliere radius does not depend on energy, the same holds for the lateral size of an electromagnetic shower, while the longitudinal development is strongly energy dependent. The lateral particle distribution is

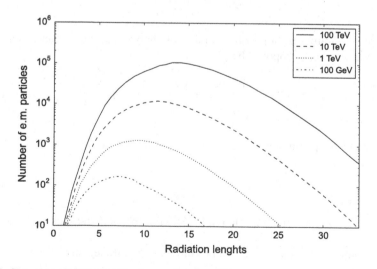

Fig. 3.4 Size of an electromagnetic shower; data from [1]

described by the *Nishimura-Kamata-Greisen* (NKG) function [7]:

$$\frac{dN_e}{rdrd\phi} = N_e(X)\frac{\Gamma(4.5-s)}{\Gamma(s)\Gamma(4.5-2s)}\frac{1}{2\pi R_M^2}\left(\frac{r}{R_M}\right)^{s-2}\left(1+\frac{r}{R_M}\right)^{s-4.5} \quad (3.15)$$

where r is the radial distance from the shower axis, Γ the Euler function.

The electrons and positrons in the shower emit Cherenkov light. The Cherenkov threshold for electrons/positrons is about 21 MeV at ground level and about 35 MeV at an altitude of 8 km. The emission angle of the Cherenkov radiation is about 1.3^0 at the the ground level, but is smaller at high altitude, steadily increasing during the evolution of the shower. The fraction of energy converted into Cherenkov light is small, but it has a very clear signature. The light arrives at the ground level as an annulus with a radius of about 120 m [11].

3.4 Interaction of Cosmic Rays: Hadronic Showers

The development of hadronic showers is governed by the nuclear interaction length λ_{int}, as the development of electromagnetic showers is characterized by the radiation length. The protons and nuclei in the cosmic rays start an hadronic shower in the first interaction length. The lateral extension of hadronic showers is larger than for electromagnetic showers, since nuclear interactions can produce particles with large transverse momentum. Primary cosmic rays with energies in the TeV region or above produce showers of secondary particles interacting with the nuclei of the atmosphere molecules, in a sequence of reactions where energetic particles can trigger the generation of new products by the same mechanism. The cascade of particles produce an *hadronic shower* [4, 10], that propagates along the direction of the primary particle The shower is not collimated, since the secondary particles have a non negligible transverse momentum and, if charged, undergo multiple scattering. The *Extensive Air Shower* (EAS) started by a cosmic rays is shown in Fig. 3.5.

We will firstly discuss the hadronic showers initiated by a proton [10]. The hadron interactions produce a large number of pions, charged and neutral, that decay through the reactions:

$$\pi^+ \to \mu^+ + \nu_\mu \quad (3.16)$$

$$\pi^- \to \mu^- + \bar{\nu}_\mu \quad (3.17)$$

$$\pi^0 \to \gamma + \gamma \quad (3.18)$$

The neutral pions are about one third of the total number of produced pions and decay into two photons, triggering the generation of electromagnetic subshowers containing photons, electrons and positrons. At each interaction, two thirds of the initial energy is released to the hadronic component, thus after k generations the energy of the hadron component will be:

Fig. 3.5 Schematic
representation of an hadronic
shower

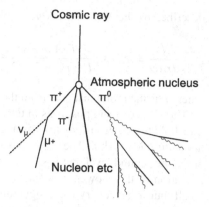

Fig. 3.5 Schematic representation of an hadronic shower

$$E_{hadron} = \left(\frac{2}{3}\right)^k E_0 \qquad (3.19)$$

The non electromagnetic component of the hadron shower is made mostly of muons. At the maximum, the energy budget of the primary is distributed between N_{max}^p photons and electrons and N_μ^p muons [6, 10]:

$$E_0 = N_{max}^p E_c + N_\mu^p E_d = g E_c \left(N_{e,max}^p + \frac{E_d}{g E_c} N_\mu^p\right) \qquad (3.20)$$

where E_d is the energy where the decay length of the pion equals the interaction length, of the order of 20 GeV.

The number of electrons at maximum and of muons is given by [6, 10]:

$$N_{e,max}^p = \left(\frac{E_0}{3 g E_c}\right) \qquad (3.21)$$

$$N_\mu^p = \left(\frac{E_0}{E_d}\right)^\beta \qquad (3.22)$$

where $\beta \sim 0.9$ [10]. The number of muons in the cascade increases almost linearly with the primary energy.

The muon component is an indicator of the hadronic energy:

$$E_h = N_\mu^p E_d \qquad (3.23)$$

The measurement of the number of electrons N_e and muons N_μ allows to reconstruct the energy of the primary, according to Eq. 3.20.

The depth of the maximum of a shower initiated by a proton is:

Fig. 3.6 The hadronic
shower front

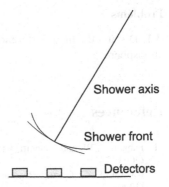

$$X^p_{max} = \lambda_{int} + X_0 \ln\left(\frac{E_0}{2 n_h E_c}\right) \tag{3.24}$$

where n_h is the number of hadrons.

The hadronic showers initiated by nuclei are discussed using the *superposition model* [10]. A nucleus with mass A and energy E_0 is equivalent to A single and independent nucleons with individual energy $\frac{E_0}{A}$, thus the hadron shower of the nucleus is the combination of A showers.

The number of electrons at maximum and the number of muons are [6]:

$$N^A_{e,max} = A\left(\frac{E_0/A}{3gE_c}\right) = N^p_{e,max} \tag{3.25}$$

$$N^A_\mu = A\left(\frac{E_0/A}{E_d}\right)^\beta = N^p_\mu A^{1-\beta} \tag{3.26}$$

The size of the electromagnetic component of hadron shower initiated by a nucleus is identical to the size of a shower started by a single nucleon. The number of muons shows a weak dependence on the mass A. The depth of the maximum of shower started by a nucleus is smaller than the depth of a shower started by a proton:

$$X^A_{max} = X^p_{max} - X_0 \ln A \tag{3.27}$$

Since the interaction length of a nucleus is $\lambda^A_{int} \sim \frac{\lambda_{int}}{A^{\frac{2}{3}}}$, the fluctuations on X^A_{max} are smaller than the fluctuations of X^p_{max}.

The structure of the hadronic shower front is shown in Fig. 3.6. The shower is a disk of particles, with a thickness of a few meters close to the axis that increases at large lateral distances. The electromagnetic and the muon components of the hadronic shower arrive at the detector at different times and can be discriminated in principle. The electrons are produced mostly in the lower parts of the atmosphere and suffer a large scattering, that increases their time of flight. The muons are produced at higher altitudes and are scattered less: the muonic front precedes the electron front.

Problems

3.1 Discuss the main differences between electromagnetic and hadronic showers in atmosphere.

References

1. Fabjan, C. W. and Gianotti, F.: Calorimetry for particle physics. Rev. Mod. Phys. **75**, 1243 (2003)
2. Gaisser, T. K.: Cosmic Rays and Particle Physics. Cambridge University Press, Cambridge (1990)
3. Greisen, K.: Cosmic ray showers. Ann. Rev. Nucl. Sci. **10**, 63 (1960)
4. Grieder, P. K. F.: Extensive Air Showers. Springer (2010)
5. Greisen, K.: End to the cosmic ray spectrum?. PRL **16**, 748 (1966); Zatsepin, G. T., Kuzmin, V. A.: Upper limit to the spectrum of cosmic rays. JETP Lett. **4**, 78 (1966)
6. Heitler, W.: The Quantum Theory of Radiation. Oxford University Press, London (1954)
7. Kamata, K., Nishimura, J.: Progr. Theor. Phys. **6** (1958) 93; Greisen, K.: Cosmic Ray Showers. Ann. Rev. Nucl. Part. Sci. **10** (1960) 63
8. Rossi, B.: High Energy Particles. Prentice-Hall, Englewood Cliffs, NJ (1952)
9. Schönfelder, V.: The Universe in Gamma Rays. Springer-Verlag Berlin Heidelberg, (2001)
10. Spurio, M.: Particles and Astrophysics - A Multi-Messenger Approach. Springer International Publishing, Switzerland (2015)
11. Weekes, T.: Very High Energy Gamma-Ray Astronomy. Institute of Physics Publishing, Bristol and Philadelphia (2003)

Part II
The World of Detectors

Chapter 4
Detectors: General Characteristics

This chapter firstly introduces the properties of detectors and presents a synoptic discussion of the organization of the following chapters. The single detectors will be the building blocks of the astroparticle instrumentation discussed later. Detectors for high energy photons and particles show a great difference compared to optical and radio detectors, the intrinsic energy discrimination capability coupled to the imaging capability. The chapter ends with a summary of some historical detectors in astroparticle physics.

4.1 Detector Properties

Detectors for particles or radiation are targeted to measure a specific physical quantity, such as the trajectory or the energy, but often they can provide more than one physical information at the same time. A large number of the detectors that will be described in the following use the physical process of ionization and the electron-ion pairs produced along the path of the charged particle [1, 2, 4]. The ionization is produced by the interaction of the projectile, but possibly also by the expelled δ-rays.

The *response function* of a detector to the radiation under investigation is the spectrum of the height of the signals measured at the detector when it is hit with monochromatic radiation. The response will not be a Dirac delta, but will have a finite width. If the detector has a linear response, the spectrum of radiation can be estimated by the distribution of pulse heights. The observed pulse height distribution $O(E)$ is the convolution of the intrinsic spectrum $I(E)$ with the response function $R(E)$:

$$O(E) = I(E) * R(E) \tag{4.1}$$

Due to different sources of fluctuations, the response function of the detector should be a Gaussian, centered at an energy E_0 and with a Full Width Half Maximum $FWHM = 2.35\sigma$. The *energy resolution* is defined as:

© Springer International Publishing Switzerland 2017
R. Poggiani, *High Energy Astrophysical Techniques*,
UNITEXT for Physics, DOI 10.1007/978-3-319-44729-2_4

$$R = \frac{FWHM}{E_0} \tag{4.2}$$

The irreducible source of fluctuations is related to the statistical nature of the detection process. The key parameter is the average energy W required to produce an information carrier, that does not necessarily coincide with the ionization potential, because of possible physical channels alternative to ionization, such as excitation. The number of produced charge carriers is:

$$n = \frac{\Delta E}{W} \tag{4.3}$$

where ΔE is the energy released in the material, W is the energy required to produce the charge carriers. The equivalent of the electron-ion pair produced in gases is the electron-hole pair in solid state materials. The energy needed to create an electron ion-pair or an electron-hole pair is of the order of 30 eV in gases and about 3 eV in solids. The statistical fluctuations of the number of produced carriers are expected to be Poissonian, $\sqrt{n}$. The observed fluctuations of the number n of charge carriers are smaller than the value expected from the Poisson statistics; the effect is described by the empirical *Fano factor F* [1, 2, 4]:

$$\frac{\sigma(n)}{n} = \sqrt{\frac{F}{n}} \tag{4.4}$$

The Fano factor is of the order of 0.05–0.2 for gases and 0.1 for solids [2].

The *efficiency* of a detector is defined as the ratio of the number of detected events to the number of incident events:

$$\eta = \frac{\# \, of \, detected \, events}{\# \, of \, incident \, events} \tag{4.5}$$

4.2 Overview of Detectors

The following chapters will present the detectors of astroparticle physics used as building blocks for ground or space based instrumentation. The main physical processes used for particle detection are:

- Ionization: the counters are based on the electron-ion pairs produced in gases and liquids or on the electron-hole pairs produced in solids by the passage of the particle
- Scintillation: the detection relies on the light flash caused by the absorption of a particle or a photon, that is transformed to an electrical signal by a photomultiplier
- Cherenkov effect
- Transition Radiation.

4.3 Historical Detectors

Some detectors have given a relevant contribution to the birth of the new sciences of cosmic ray astrophysics and of high energy physics and are described here for completeness [1, 3]. The historical detectors are not triggerable as the detectors described in the following chapters.

The *cloud chamber* has been the work horse of high energy astrophysics at its infancy, allowing the discovery of the positron and the muon in cosmic rays. A cloud chamber is a vessel containing a mixture of gas and vapor operating at the pressure of saturation of the vapor. The passage of a charged particle produces a ionization track, with a lifetime of the order of several millisecond. The start of adiabatic expansion decreases the temperature of the mixture and supersaturates the vapor component, that condenses on the positive ions along the track forming small droplets within hundreds of milliseconds. The droplets are finally photographed. The cloud chamber must be cleaned of all positive ions and undergo the compression of the mixture before starting a new measurement.

The *nuclear emulsions* are made of AgBr crystals inside a gelatin substrate. The emulsions allow the reconstruction of the trajectory of charged particles: the extraction of charges due to the ionization energy loss transforms silver bromide into silver. The development and fixation stages produce an image of the ionization profile of the charge. The typical thickness of an emulsion is of the order of hundreds microns. The emulsion grains should be small to achieve a high spatial resolution: the typical grain size is of the order of 0.1–1 μm. The emulsions have been extensively used for cosmic ray experiments in the past. The radiation length of typical emulsions is of the order of a few centimeters, thus they can be used to detect electromagnetic cascades. X-ray films are a family of emulsion with smaller grains (by an order of magnitude) and a thickness of the order of tens microns.

The *plastic detectors* are based on the destructive effect of the passage of a highly charged particle on the physical structure of the material. The damaged section are detected by using an etching compound that preferentially reacts with the undamaged material and leaves etch cones as markers of the trajectory. The amount of damage to the material is proportional to the square of the particle charge and to a function of its velocity. The energy loss is deduced by the size of the cones left by the etching process. If an independent measurement of the particle velocity is available, the particle charge can be estimated. The plastic detectors have been used on board of balloons to measure the composition of cosmic rays.

The *spark chamber* was used to detect photons with an energy above some tens MeV. A spark chamber consists of a set of converter plates inside a sealed vessel containing gas. The plates are made of a material with high atomic number (to have a high conversion probability) that are alternatively fed by high voltage or grounded. Two scintillators are placed above and below the spark chamber. The passage of a particle activates the scintillators that trigger the high voltage input to the metal plates. A spark develops along the track of the particle and is recorded photographically or electronically. The reconstruction of the track in three dimensions is performed using two orthogonal readout systems. The produced charges must be removed with a suitable clearing field to allow another detection.

Problems

4.1 Discuss the main features of emulsions.

4.2 Discuss the main features of spark chambers.

References

1. Grupen, C. and Swartz, B.: Particle Detectors. Cambridge University Press (2008)
2. Leo, W. R.: Techniques for Nuclear and Particle Physics Experiments - A How-to Approach. Springer-Verlag (1994)
3. Longair, M. S.: High Energy Astrophysics: Volume 1. Photons, Particles and their Detection. Cambridge University Press (1992)
4. Olive, K.A. et al. (Particle Data Group): Chin. Phys. C **38**, 090001 (2014)

Chapter 5
Detectors Based on Ionization in Gases and Liquids

This chapter presents the detectors based on ionization in gases and liquids. Ionization detectors are discussed starting from the ionization in gases, the basis of several detectors of this class: the incident particles produces electron-ion pairs that undergo avalanche multiplication in high electric fields. Ionization chambers, proportional counters and Geiger counters are based on the above principle, but are operating at different electric fields. The evolution of these systems has lead to the building of Multi Wire Proportional Chambers and drift chambers. Liquid ionization detectors, that rely on the higher density of the ionizing medium, are discussed. The energy resolution capability of gas and liquid based detectors is presented.

5.1 Ionization in Gases

A charged particle passing through a gas in a container produces electron-ion pairs by ionization along its track [1, 5, 6]. The energy required to create a pair is of the order of 30 eV. A neutral particle needs an additional step, a preliminary interaction that releases charges triggering the ionization process. The *ionization chamber* separates electrons and ions with an electric field directing them towards the anode and the cathode. The simplest arrangement consists of an anode wire and of an external shell cathode (Fig. 5.1), with a radial electric field behaving as $\frac{1}{r}$, where r is the distance from the anode wire. The filling medium is a noble gas (Xe, Ar, Kr, He, Ne) or a mixture of gases (Ar with ether, ethanol, acetone).

The number of ions collected in the detector depends on the voltage applied to the electrodes (Fig. 5.2), that is of the order of hundreds Volts at least. The ionization chamber, the proportional chamber and the Geiger counter are different manifestations of the same physical process. At low voltages, only a few electron-ion pairs are generated, but undergo recombination and are not detected. With higher voltages, some charges will be collected, until all charges will be collected before recombining: this is the regime of *ionization chamber*, independent from the applied voltage.

© Springer International Publishing Switzerland 2017
R. Poggiani, *High Energy Astrophysical Techniques*,
UNITEXT for Physics, DOI 10.1007/978-3-319-44729-2_5

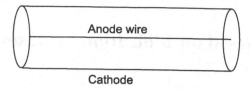

Fig. 5.1 Cylindrical ionization detector

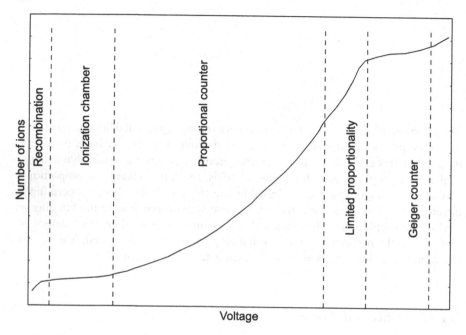

Fig. 5.2 Number of ions collected in a ionization gas detector

A further increase of voltage produces a linear increase in the number of collected ions, since the electric field accelerates the ionization electrons that produce secondary ionization events and, finally, an avalanche. The multiplication of particles is concentrated close the anode wire. The number of pairs in the avalanche is proportional to the number of primary electrons produced by ionization: this if the region of *proportional chamber* operation, where the gain factor in the number of collected charges can exceed 10^5. The linear response makes this regime the most used for detection. The linearity is gradually lost at larger voltages, since the charge in the gas is large enough to distort the electric field at the anode. At larger voltages, a discharge is triggered inside the gas along the anode wire length: the chamber signal is large, but saturated, and the discharge must be reset by using quencher gases.

Table 5.1 Properties of some gases at NTP [6]

Gas	W_I (eV)	$\frac{dE}{dx}$ (keV cm^{-1})	N_{pair} (cm^{-1})	N_{total} (cm^{-1})
He	41.3	0.32	3.5	8
Ar	26	2.53	25	97
iC$_4$H$_{10}$	26	5.67	90	220
CF$_4$	54	6.38	63	120

The properties of some gases at normal temperature and pressure are shown in Table 5.1: the average energy W_I per ion pair, the differential energy loss $\frac{dE}{dx}$, the number of primary (N_{pair}) and total (N_{total}) electron-ion pairs per cm per a minimum ionizing particle with unitary charge.

5.2 Proportional Counters

The proportional counters [1–6, 8] are very used in high energy astrophysics to detect X-rays. The dominant absorption mechanism is the photoelectric effect, up to energies of the order of tens keV. The filling gases are noble gases with high atomic number, since the cross section of the photoelectric effect is proportional to Z^5. The number of electrons generated by the absorption is governed by the average energy required to create a pair, of the order of 30 eV: thus a 3 keV photon will produce about one hundred of electron-ion pairs. The electrons and the ions drift towards the anode and the cathode, respectively. At high electric fields, the electrons extracted in the primary ionization can gain enough energy to trigger secondary ionization and increase the number of charges that are detected, providing an *ionization gain* through the formation of an avalanche. In a cylindrical detector, the avalanche is formed in the region close to the anode wire, where the electric field can be as high as a few tens kV cm^{-1}. The moving charges induce currents on the anode and the cathode. Due to the higher mobility of the electrons, about 10^2 cm^2/V s, one thousand times larger than the mobility of ions, the output signal contains a fast rising contribution produced by the electrons and a slower contribution caused by the ions. Since the larger fraction of the ion pulse is related to the initial part of the drift towards the cathode, the use of an *RC* differentiator allows to have a fast signal at the output. The avalanche is localized close to the point of production. The time resolution can be of the order of 1 μs or below.

The total number of ionization events per unit length α is called the *first Townsend coefficient* [1, 5, 6]:

$$\frac{\alpha}{p} = A e^{-\frac{B}{\frac{E}{p}}}$$

(5.1)

where p is the pressure, A, B are constant coefficients. The magnitude of the quantity $\frac{\alpha}{p}$ for noble gases is of the order of 1 pair/cm mm(Hg) for values of $\frac{E}{p}$ of about 10^2 V/cm/mm(Hg) [1, 6].

The number of particles at a position x is:

$$N(x) = N_0 e^{\int \alpha(x)dx} \tag{5.2}$$

The amplification factor is:

$$G = e^{\int \alpha(x)dx} \tag{5.3}$$

The integration is performed from the position where charge multiplication starts to the anode radius. In the proportional regime of operation, the amplification factor is constant and the output signal is proportional to the ionization. Typical gain factors are of the order of 10^4 to 10^6 [1, 6].

The electrons in the avalanche produce also excitation of the gas molecules, in addition to ionization. The deexcitation process occurs through emission of ultraviolet photons, that can inject additional electrons by undergoing photoelectric effect on the cathode surface. To avoid discharges, the ultraviolet photons are absorbed by adding *quench gases* to the detector gas. A standard recipe for a proportional counter for X-rays is the filling with a mixture of noble gases (mainly argon or xenon) with the addition of quenchers (up to several percent).

The efficiency of a proportional counter depends on the gas used as the ionization medium, but also on the properties of the entrance window. The window should be thin enough to avoid relevant absorption and strong enough to withstand the gas pressure. The windows are made of metal (typically beryllium) or plastic materials. The former material allows the detection of X-rays above a few keV, while the latter allows the detection of X-rays with smaller energies, but the intrinsic permeability of the material requires a refilling of the gas.

The proportional counters have the capability of intrinsic energy resolution. The resolution is determined by the statistics of the initial ionization event and by fluctuations of the avalanche process [1, 3, 8]. The first contribution is described by the variance $(\frac{\sigma_n}{n})^2 = \frac{F}{n}$, where F is the Fano factor. The second contribution is related to the fluctuations of the amplification A, according to $(\frac{\sigma_A}{A})^2 = b$. The energy resolution is thus given by:

$$\frac{\sigma_E}{E} = \sqrt{\frac{F+b}{n}} = \sqrt{\frac{W(F+b)}{E}} \tag{5.4}$$

where W is the energy required to create an electron-hole pair, about 30 eV. The Fano factor F ranges from 0.05 to 0.3 and the b constant related to multiplication is in the interval from 0.5 to 0.8. As an example, for CH_4 the Fano factor is 0.26, while b is 0.75 [9].

The operation of proportional counters in space has to face the background of cosmic rays, that has a rate larger than the rate of the typical astrophysical X-ray sources; in addition, cosmic rays release energies that are three orders of magnitude at least larger than the X-ray signals. The rejection of the background is achieved by the information in the morphology of the avalanche. Since the avalanche triggered by X-rays is more localized than the ionization triggered by a cosmic ray, that is more similar to a track, the events can be discriminated by the difference in the rise time of the signal. The large energy release of cosmic rays in the gas can induce a discharge in the gas. The lifetime of proportional counter in space is limited by the radiation environment. The quencher molecules are cracked and produce polymers that deposit on the anode and cathode. The detector must be equipped with a system for background rejection, as will be discussed in Chap. 11.

The low fluxes of astrophysical X-ray sources has triggered the development of proportional counters with large collection areas, of the order of thousands cm^2, with fields of view of the order of few square degrees [8]. The instruments have a cell structure, with multi anode wires and cathode grids. The signal of genuine X-rays and of charged particles can be discriminated by setting a veto on the cells surrounding the cell with the candidate event.

The proportional counters have an intrinsic spatial resolution of the order of 50–100 μm [6]. The time resolution is limited by the drift velocity of electrons and is of the order of a few nanoseconds [6].

5.3 Geiger-Muller Counters

Geiger-Muller counters [1, 5, 6] are operated at electric fields high enough to produce several photons during the charge multiplication process: the photons produce additional charges by photoelectric effect, not necessarily close to the origin of the primary avalanche [1, 5]. A discharge develops along the anode wire; the signal will depend only on the bias voltage, but no more on the primary ionization. The number of ions can be as high as 10^{10}. During the discharge, the detector is not sensitive to other incident particles. The discharge is stopped by lowering the anode voltage below the threshold for Geiger operation, until the collection of ions at the cathode is completed. The required time is of the order of milliseconds, thus the Geiger-Muller detectors cannot be used in high rate environments. An alternative solution is the use of quenchers (at the level of several percent) to absorb the photons produced during the multiplication and reduce their contribution.

5.4 Streamer Tubes

Streamer tubes [1, 5, 6] rely on the generation of avalanches close to the anode wire, but use a high content of quencher to limit the extension of the discharge along it [1]. The mechanisms produces a high number of photons that are promptly absorbed close to the original position to produce additional charge avalanches. Since the avalanche is localized, the streamer tubes are able to detect multiple particle at the same time. The operation of streamer tubes requires high voltages, of the order of a few kV. If the anode wire is thicker than in the proportional chamber, with a diameter of some tens microns, the avalanche will be very close to it and the signal can be extracted from the anode.

5.5 Multi-Wire Proportional Chambers

The *Multi-Wire Proportional Chamber* (MWPC) consists of a set of proportional counters with parallel anode wires that share the same gas filled enclosure (Fig. 5.3) [1, 3, 5, 6, 8]. The typical diameter of anode wires is of the order of a few tens microns, with a wire spacing of a few millimeters and a distance to the cathode of the order of centimeters. The gas filling the MWPC are the same gases used in proportional counters. The wire spacing is limited by the electrostatic repulsion between the wires, that determines the tension required for their stable operation, and by the gravitational sag produced by their own mass.

Assuming that the anode wires are positioned at the coordinates $x = 0, \pm d$, $\pm 2d, ...$ and $y = 0$, the potential distribution is given by:

$$U(x, y) = \frac{CV}{4\pi \epsilon_0} \left[2\pi \frac{L}{d} + \ln \left[4 \left(\sin^2 \frac{\pi x}{d} + \sinh^2 \frac{\pi y}{d} \right) \right] \right] \qquad (5.5)$$

where d is the distance between wires, L is the distance of wires from the cathode plane, C the capacitance per unit length, V is the anode voltage. The electric field near each anode wire is close to the field of the single anode wire of the proportional counter and the avalanche formation is similar. The MWPC can achieve a gain factor of the order of 10^5.

The MWPC provides the coordinate orthogonal to the wires. The position of an event can be reconstructed by dividing a cathode into parallel strips and measuring the signals induced on it. The coordinate along the wire is extracted from the center of gravity of the charges. The reconstruction is improved by segmenting both cathodes in orthogonal directions. The spatial resolution is of the order of some tens μm [6].

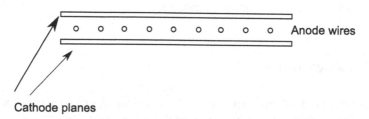

Fig. 5.3 Layout of a Multi-Wire Proportional Chamber

Fig. 5.4 Layout of a drift
chamber

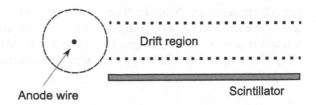

The MWPCs used for X-ray astrophysics use a modification of the original design
[3, 8]. The grid of anode wires is mounted between two cathode wire planes arranged
as grids to provide position sensitivity. An additional anode grid is mounted below
one cathode grid to provide an anticoincidence signal for background rejection. The
X-rays enter the MWPC through a window, cross the first cathode grid and enter the
region of the anode wire where they undergo multiplication.

5.6 Drift Chambers

A *drift chamber* uses a set of anode wires inside two cathode planes as the MWPC
[1, 5, 7]. The position of a track is estimated by the measurement of the arrival time
of the electrons produced by the passage of the particle to the anode (Fig. 5.4).

If v_d is the drift velocity of the electrons in the gas, that is of the order of some
centimeters per microsecond, then the distance of the interaction point from the anode
is:

$$x = \int v_d dt \tag{5.6}$$

The electric field for the drifting of the charges is provided by a set of additional
wires. The addition of a grid separates the gas amplification from the drift process.

The measurement of the drift time allows to use a smaller number of wires compared to the MWPC. The spatial resolution is determined by the time resolution of
the instrumentation measuring the drift time. With a drift velocity of some cm/μs
and a time resolution of 2 ns, the spatial resolution is of the order of 80 μm.

The concept of the drift chamber has been extended to the *Time Projection Chamber* (TPC), that provides a measurement of the drift time and of the charge induced
on the cathodes.

5.7 Microstrip Gas Chambers

The *Microstrip Gas Chambers* addresses the issues of stability of the wires in the
MWPC systems [1, 6]. The anodes and cathodes are manufactured as a set of con-
ductive microstructures on a substrate using electron lithography. The anode and

cathode elements are alternated. The compact structure produces high electric field around the anode strips to amplify the charges produced by photoionization and a fast elimination of the ion charges. The detector has two dimensional position capability. The amplification factor is comparable with that of the classical MWPC, as the spatial resolution.

5.8 Liquid Ionization Detectors

A ionization chamber using a liquid as ionizing medium has a larger absorption efficiency compared to one using a gas or a mixture of gases, due to the much larger density [1, 6]. The filling liquids are liquefied noble gases operating at cryogenic temperatures: liquid argon (85 K), xenon (163 K), krypton (117 K). The ionization potential is in the range 15–25 eV. The liquids must have a high purity level, due to the low drift velocity of charges in the liquid. A drawback of the liquid ionization detectors is the necessity of operating the electronic readout at cryogenic temperatures.

Problems

5.1 Discuss the different types of gas ionization detectors.

5.2 Discuss the problems related to the operation of gas ionization detectors in space.

References

1. Grupen, C. and Swartz, B.: Particle Detectors. Cambridge University Press (2008)
2. Grupen, C. and Buvat, I: Handbook of Particle Detection and Imaging. Springer-Verlag Berlin Heidelberg (2012)
3. Huber, M. C. E., Pauluhn, A., Culhane, J. L., Gethyn T. J., Wilhelm, K., Zehnder, A.: Observing Photons in Space - A Guide to Experimental Space Astronomy. Springer Science+Business Media, New York (2013)
4. Lèna, P. et al.: Observational Astrophysics. Springer-Verlag Berlin Heidelberg (2012)
5. Leo, W. R.: Techniques for Nuclear and Particle Physics Experiments - A How-to Approach. Springer-Verlag (1994)
6. Olive, K.A. et al. (Particle Data Group): Chin. Phys. C **38**, 090001 (2014)
7. Schoönfelder, V.: The Universe in Gamma Rays. Springer-Verlag Berlin Heidelberg, (2001)
8. Trumpër, J.E. and Hasinger, G.: The Universe in X-Rays, Springer-Verlag Berlin Heidelberg (2008)
9. Zombeck, M.V.: Handbook of Space Astronomy and Astrophysics. Cambridge University Press (2007)

Chapter 6
Scintillation Detector Systems

This chapter discusses the scintillators, materials that produce small amounts of light when hit by radiation. The light emitted by scintillators is collected by photomultiplier tubes, light detectors with an high internal gain. Scintillators are made of different materials (inorganic, organic, gas, liquids) with different light emission mechanisms. The scintillators of interest for astroparticle physics are presented.

6.1 Scintillators

Scintillation detectors or *scintillators* are based on the emission of a light pulse after the absorption of radiation or particles [1–7]. The pulse is converted into an electrical signal by a *photomultiplier*. The process is called *luminescence*: the energy of the incident particle is absorbed and reemitted within some tens of nanoseconds. Scintillators have several advantages. Generally they have a *linear response* in energy and a fast *temporal response*. For several of them, the shape of the light pulse depends on the type of the incident particle or radiation, allowing its identification. Usually the light pulse shows a very fast rise, followed by an exponential decay with a typical decay time (or, more rarely, by a decay with a fast and a slow component).

There are several classes of scintillators, classified according to the material type. A synoptic view of the properties of typical scintillators that will be discussed below is reported in Table 6.1. The performances of a scintillator are summarized by the light yield, the number of photons emitted by the material per MeV of absorbed energy. The emission spectrum has a maximum, whose wavelength determines the characteristics of the photon transducer. Generally the peak value is in the optical range and matched to the photomultiplier spectral sensitivity, that has a maximum at about 400 nm. Scintillators with peak emission in the ultraviolet are doped with *wavelength shifters* to shift the peak in the optical range.

The family of *inorganic crystals* includes undoped crystals and alkali halides doped with impurities that are responsible for the luminescence. In both cases the

© Springer International Publishing Switzerland 2017
R. Poggiani, *High Energy Astrophysical Techniques*,
UNITEXT for Physics, DOI 10.1007/978-3-319-44729-2_6

Table 6.1 Scintillator type, material, density, decay time, peak wavelength, light yield (photoelectrons per MeV); data from [3, 5]

Type	Material	Density (g/cm³)	Decay time (ns)	Peak wavelength (nm)	Light yield (photons/MeV)
Inorganic	NaI(Tl)	3.67	230	415	37,700
Inorganic	CsI(Tl)	4.51	600, 3400	550	64,800
Inorganic	CsI(Na)	4.51	630	420	38,500
Inorganic	BGO	7.13	300	480	8200
Inorganic	LSO	7.40	40	420	30,000
Inorganic	GSO	6.71	56, 600	440	12,500
Inorganic	BaF_2	4.88	0.8, 630	220, 315	1400, 10,000
Inorganic	$PbWO_4$	8.28	10, 30	420	100, 31
Organic	Anthracene	1.25	3.0	445	20,000
Plastic	Bicron BC-408	1.03	2.1	425	10,000
Plastic	UPS-89	1.06	2.4	418	10,000
Liquid	Hydrocarbon	0.88–1.0	1–300	425	12,000

scintillation is related to the band structure of the material. The valence band and the conduction band are separated by an energy gap of the order of a few eV. The former is filled, while the latter is empty. The energy released by the incident radiation lifts the electrons in the conduction band, leaving holes in the valence band. An electron can undergo recombination with an hole or form a bound state with it, named *exciton*, whose energy level is just below the bottom of the conduction band. The exciton moves inside a the crystal, ending with disexcitation or recombining with the emission of one photon, with a probability increasing with decreasing temperature. The addition of *impurities* in the crystal creates an additional energy band intermediate between the valence and the conduction band. Electrons and excitons can transfer energy to the impurities, that release it by photon emission or as lattice vibrations. Inorganic crystals have high densities and an high atomic number, thus they are suitable for detection of gamma rays. The most used materials are NaI(Tl), CsI(Tl), CsI(Na), BGO, BaF_2. NaI(Tl) and CsI(Na) have an emission spectrum well matched to the photomultiplier spectral input. The light yield of crystal scintillator is large, of the order of 10^3 to 10^4 photons per MeV: some tens of eV are needed for the emission of a photon. CsI(Tl) has the highest light yield, but is not matched to the spectral response of the photomultipliers and requires other photon transducers, such as photodiodes. The time response of inorganic scintillators is of the order of hundreds of nanoseconds or microseconds, slower than that of other scintillators. Some inorganic crystals (NaI(Tl), CsI(Na)) are hygroscopic and must be operated inside sealed containers. The inorganic crystals, NaI(Tl), CsI(Tl), CsI(Na), are the suitable choice for X-ray and γ ray detectors, due to their high atomic number.

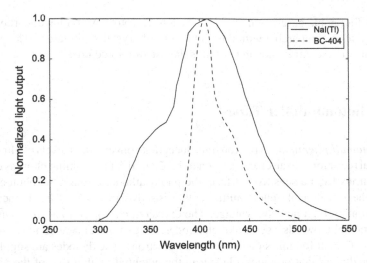

Fig. 6.1 Emission spectra of the inorganic crystal NaI(Tl) and of the plastic scintillator BC-404; data from Bicron catalog

The *organic scintillators* include a few crystals and a large variety of plastics and liquids. In the last two cases the scintillator material is a solute, dissolved in an organic solid or a liquid medium. Plastic scintillators are made of polymeric materials with a chemical structure containing benzene rings. The most common materials are polystyrene and polyvinyltoluene, that can be produced in large volumes and are easily machinable to any desired shape. The photon emission of plastic scintillators is triggered by the passage of charged particles. The main organic crystal is anthracene. The light output of organic scintillators is lower than that of inorganic crystals. On the other hand, the decay time of the scintillation light is very fast, of the order of nanoseconds, thus plastic scintillators are very suitable for time of flight measurements.

The response of organic scintillators is almost linear, with a deviation at high ionization density described by the *Birks law*:

$$Y = Y_0 \frac{1}{1 + k_B \frac{dE}{dx}} \tag{6.1}$$

where Y is the light yield, $\frac{dE}{dx}$ is the energy loss by ionization, k_B is the Birks' coefficient, of the order of a few time 10^{-3} g/(cm^2 MeV).

The *gas scintillators* produce light during the disexcitation following the absorption of the energy of an incident particle.

The emission spectra of two typical scintillating materials, the crystal NaI(Tl) and the plastic BC-404, are reported in Fig. 6.1 for reference.

Liquefied noble gases such as liquid Xenon, Krypton, Argon have a probability of scintillation comparable with the probability of crystal scintillators. The peak of emission is in the ultraviolet and requires the use of photodiodes.

6.2 Photomultiplier Tubes

The *photomultiplier tube* is a photon transducer that converts faint levels of ultraviolet or optical radiation into an electrical signals [1, 2, 4, 5]. The incoming photons extract the electrons from a sensitive surface, the photocathode, through the photoelectric effect. The operation of a photomultiplier tube is shown in Fig. 6.2. The photocathode is biased at an high negative voltage. The extracted photoelectrons are focused by an electric field towards a system of electrodes, the dynodes, made of materials with a high coefficient for emission of secondary electrons. The dynodes are supplied by a voltage divider that provides a ladder in the potential, with a step of the order of 100 V. The initial charge extracted from the photocathode is amplified by a large factor before arriving to the anode. The whole assembly is housed inside a vacuum evacuated enclosure.

The quantum efficiency of the photocathode, i.e. the ratio of the number of photoelectrons produced for each incoming photon, is of the order of 15–25 % (Fig. 6.3).

The amplification factor of a photomultiplier tube with m dynodes with secondary emission coefficient k is $G = k^m$. Assuming $k = 3$, $m = 14$, the gain factor is $\sim 4.8 \cdot 10^6$, thus the charge at the anode will be $\sim 7.7 \cdot 10^{-13}$ C; if it is collected within 4 ns, the current at the anode will be of the order of 0.2 mA, that, on a 50 Ω resistor will produce a voltage of about 10 mV, that can be measured very easily. Typical gains range from 10^3 to 10^7. The photomultiplier tubes are very fast devices, with a rise time of the order of a few nanoseconds. The transit time of electrons inside the device is of the order of tens nanoseconds. The time resolution is limited by the variation in the transit time and by the time jitter [4]. The transit time spread depends on the energy and the direction of emission. Usually it is of the order of a fraction of nanosecond, but it can become of the order of a few nanoseconds in large photomultiplier tubes like those used for neutrino experiments (Chap. 14). The time

Fig. 6.2 Layout of a photomuliplier tube, adapted from http://www. hamamatsu.com/jp/en/ technology/innovation/ photocathode/index.html

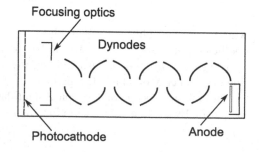

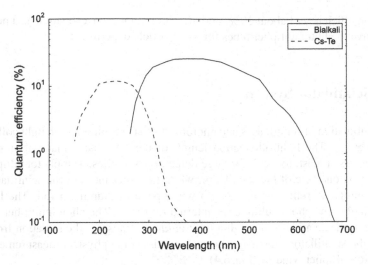

Fig. 6.3 Quantum efficiency of different photocathode materials; data from http://www.hamamatsu.com/jp/en/technology/innovation/photocathode/index.html

jitter is caused by the fact that the photoelectric effect and the charge multiplication are statistical processes.

The photomultiplier tubes must be shielded from external magnetic fields, including the Earth magnetic field, with shields of high permeability material, the μ-metal. In addition, they should operate at a stable temperature.

The photomultiplier tubes are extensively used in ground based gamma ray astronomy to measure the Cherenkov light emitted by the charged particles of the showers.

The *Micro-Channel Plates* (MCP) [3, 7] are a different implementation of the photoelectric effect. The MCP is made of many small glass tubes internally coated with a photocathode material and fused together. The tube system is positioned inside a high electric field. The incident radiation produces photoelectrons that trigger an avalanche as in a photomultiplier; the charge is collected with anode wires. The MCPs are used in ultraviolet and X-ray astronomy.

There are some other technological solutions for the detection of faint levels of light. The *photodiodes* (PD) are used as light detectors in alternative to the photomultiplier tubes for some applications [5]. A photodiode is a reverse biased *pn* junction. Photons with wavelengths smaller than about 1000 nm produce electron-hole pairs by the photoconductive effect; the efficiency is of the order of 90 %. In the *avalanche photodiodes* the initial pair generates a cascade of impact ionization events. Both families of detectors have been used for the readout of the light of crystal scintillators in calorimeters. The *Silicon Based PhotoMultipliers* (SiPM) are APD detectors operating in Geiger mode, where a photoelectron starts an avalanche breakdown. The output does not depend on the number of primary charges. The SiPM have high gains, of the order of 10^5, at bias voltages of the order of some tens Volt and a photon detection efficiency of about 20%. and are current under investigation as a possible alternative to photomultiplier tubes for astroparticle experiments.

6.3   Scintillator Systems

The combination of scintillators and photomultipliers requires a high light collection efficiency [4]. The light attenuation length of the scintillator material should be some meters at least, to use it for large detector assemblies. The photomultiplier is mounted on one face of the scintillator, while the other faces of the scintillator are made completely reflecting by using a white paint or aluminum foil. The light is transmitted to the photomultiplier by internal reflection. The photomultiplier can be coupled to the scintillator through a *light guide*, to transport the radiation from the face of the scintillator, whose geometry depends on the physical measurements, to the photomultiplier window (Fig. 6.4).

The technique of *Pulse Shape Discrimination* (PSD) allows to discriminate between different incident particles by the shape of the scintillation pulse, whose decay time depends on the type of particle. For example, CsI(Tl) has a decay time of about 0.52 μs for protons and of about 0.70 μs for electrons. The operation of scintillation systems for gamma ray observatories on board of satellites has to deal with the background of cosmic rays (Chap. 13). The *phoswich* technique combines two scintillator with different decay times. The scintillator devoted to the detection is shielded with the second scintillator. The two scintillators are optically coupled and are read by the same photomultiplier tube. The technique of Pulse Shape Discrimination allows to reject the events occurring in the shielding scintillator.

The scintillator-photomultiplier systems are often used for gamma ray spectroscopy. The spectrum of the gamma rays emitted by a ^{137}Cs radioactive source measured with a NaI(Tl) scintillator is shown in Fig. 6.5. The rightmost feature is the *photopeak*, related to the complete absorption of the gamma rays. The small peak at intermediate energy is the *Compton edge*, the cut-off of the Compton continuum. The leftmost peak is caused by the *back-scatter* of photons that scatter on the material around the detector and return to it. The energy resolution achievable in scintillator based gamma ray spectrometers is of the order of 10%.

Fig. 6.4 Scintillator coupling to the photomultiplier through a light guide

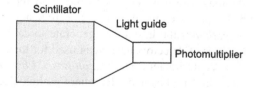

Fig. 6.5 Energy spectrum of a ^{137}Cs radioactive source measured with a NaI(Tl) scintillator

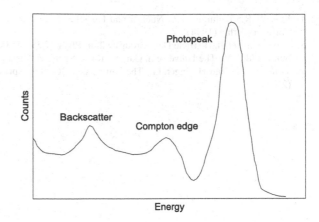

6.4   Gas Scintillation Proportional Counters

The *Gas Scintillation Proportional Counter* (GSPC) [3, 7] combines the features of the ionization energy loss with those of the scintillation. The fluctuations of the number of ionizing events during the avalanche deteriorates the energy resolution of a standard proportional counter. The absorption of the X-ray occurs in a region with low electric field. The produced charges are not amplified, but they drift into an high field region where they gain enough energy to produce scintillation in the gas. The scintillation photons, typically ultraviolet, are detected with a photomultiplier. The number of scintillation photons is proportional to the number of interactions of the electrons with the gas atoms. The energy resolution is governed by the fluctuations of the scintillation photons and is better than in standard proportional counters, approaching the Fano factor. The event position can be localized by the estimation of the centroid of the ultraviolet light, replacing the end photomultiplier with an array of photomultipliers or a multianode photomultiplier.

Problems

6.1 Discuss the different types of scintillators and their application.

6.2 Discuss the main properties of photomultiplier tubes.

References

1. Grupen, C. and Swartz, B.: Particle Detectors. Cambridge University Press (2008)
2. Grupen, C. and Buvat, I: Handbook of Particle Detection and Imaging. Springer-Verlag Berlin Heidelberg (2012)
3. Huber, M. C. E., Pauluhn, A., Culhane, J. L., Gethyn T. J., Wilhelm, K., Zehnder, A.: Observing Photons in Space - A Guide to Experimental Space Astronomy. Springer Science+Business Media, New York (2013)

4. Leo, W. R.: Techniques for Nuclear and Particle Physics Experiments - A How-to Approach. Springer-Verlag (1994)
5. Olive, K.A. et al. (Particle Data Group): Chin. Phys. C **38**, 090001 (2014)
6. Schönfelder, V.: The Universe in Gamma Rays. Springer-Verlag Berlin Heidelberg (2001)
7. Trumpër, J.E. and Hasinger, G.: The Universe in X-Rays, Springer-Verlag Berlin Heidelberg, (2008)

Chapter 7
Detectors Based on Ionization in Solid State Materials

Solid state detectors are ionization detectors with a solid as the detecting medium. These detectors combine high energy resolution with high spatial resolution. Charged particles or photons produce electron-hole pairs, requiring a smaller energy to create the pair compared to the energy required to create a pair in gases or produce a light signal in scintillators. Solid state detectors are practical solutions for space based observatories, since they can provide imaging and spectroscopic capabilities within small volumes. Detectors based on high purity semiconductors operating at low temperatures will be presented. The space experiments use semiconductor based detectors with high spatial resolution, silicon strip detectors, as the ones used for tracking at colliders. The Charge Coupled Devices, the standard detectors of optical astronomy, are discussed as imaging and spectroscopic detectors for X-ray photons.

7.1 Material Properties

The solid state detectors are ionization chambers with the gas replaced by a solid material [1–8]. The absorption of a photon or a particle produces an electron-hole pair, in analogy to the electron-ion pair discussed before. Since the density of solids is much higher than that of gases, they have a better absorption probability and can be used at higher energies. The high density allows to build compact instruments. The most used solid materials are semiconductors, that have a band gap energy of the order of a few eV. The solid state detectors have a better energy resolution, compared to scintillators or gas ionization detectors, where energies of the order of 100 or 30 eV are requires to generate the information carriers. The number of charge carrier is larger by one or two orders of magnitude, thus the energy resolution is improved. The solid state detectors can be manufactured with thin strips to produce trackers with high spatial resolution and low mass, suitable for satellite observatories.

© Springer International Publishing Switzerland 2017
R. Poggiani, *High Energy Astrophysical Techniques*,
UNITEXT for Physics, DOI 10.1007/978-3-319-44729-2_7

Table 7.1 Physical properties of some semiconductor materials; data from [3]

Property	Si	Ge	CdTe	HgI$_2$
Z	14	32	48/52	80/53
ε	11.9	16	10.36	8.8
Band gap (eV)	1.12	0.68	1.52	2.13
Ionization energy (eV/pair)	3.61	2.98	4.43	4.3

The properties of the main solid state materials are summarized in Table 7.1.

The Fano factor of semiconductor detectors is of the order of 0.1. Assuming a band gap of 1 eV, the energy resolution that can be potentially achieved with semiconductor detectors is of the order of 10^{-3}, superior to the resolution achieved by scintillator based systems, of the order of several percent.

The basic component of a semiconductor based detector is the *pn* junction, with its *depletion region*. The application of a reverse bias increases the thickness of the depletion layer. The charges produced by an incident particle or photon are separated by the electric field and produce a current in the detector. The total charge collected at the electrodes is proportional to the absorbed energy. The thickness of the depletion region is:

$$d_d = \sqrt{2\varepsilon\mu\rho(V_b + V)} \tag{7.1}$$

where ε is the dielectric constant of the material, ρ the resistivity, μ the mobility of charge carrier, qV_b is the energy difference of the two sides of the junction, V is the external bias voltage. The semiconductor detectors do not have an internal gain mechanism, thus are equipped with charge sensitive preamplifiers. The mobility of electrons and holes is very similar, of the order of 10^3 cm^2/V s, thus both of them contribute to the final signal. Semiconductor detectors with thick depletion layers are largely used for X-ray detection.

The intrinsic energy resolution of semiconductor materials makes them suitable for X-ray and gamma ray spectroscopy. The high atomic number increases the probability of the photon interaction. The absorption coefficients of some materials are reported in Fig. 7.1.

Silicon is used in the X-ray region, due to its low atomic number. The detection of gamma rays in the MeV region requires a large depletion depth. Silicon (and germanium) can achieve high resistivity using the technique of *compensation*. The addition of lithium, an electron donor, into the *p* layer, produces a region where the doping is compensated. The sensitive layer has a thickness of a few millimeters. The modified materials are called Si(Li) and Ge(Li) and have a resistivity close to that of undoped semiconductors. The new detector structure is named *pin*, where *i* is the label for intrinsic. The compensated detectors must be maintained at low temperatures also when they are not in operation to block the rediffusion of the lithium. The *High purity semiconductors* [1–5, 7, 8] have very small impurity concentrations, of the

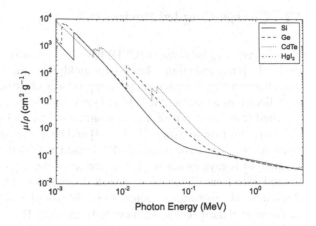

Fig. 7.1 Absorption coefficient of semiconductor materials (dat rom NIST)

Fig. 7.2 Layout of a silicon microstrip detector

order of 10^{10} cm^{-3}. The thickness of the sensitive region can be as large as a few centimeters. The detectors are cooled at the liquid nitrogen temperature for operation but it is not necessary to keep them cooled when not in operation. The most common detector is the High Purity Germanium (HPGe) detector, a traditional choice for gamma ray spectroscopy. Additional detectors use compound semiconductors that are able to operate at room temperature due to their large band gap: CdTe, HgI$_2$, CdZnTe. The CdTe, CdZnTe and HgI$_2$ have found a large number of applications in small scale gamma ray spectrometers and have been used on several satellite missions. The detection of UV radiation in space has triggered the development of solar blind detectors, with wide band gaps [3] and cut-off wavelengths of about 200 nm. The main materials are diamond and Al$_x$Ga$_{1-x}$N with different element ratios. The solar blind detectors have small dark currents.

One of the main applications of solid state detectors is the tracking of the trajectory of particles. The electrodes are manufactured with different morphologies, for examples as microstrips or pads, to provide the reconstruction of the position (Fig. 7.2).

The silicon strip detectors used in astronomy are modeled on the microstrip detectors used for high energy physics experiments [1]. The distance between strips is of the order of 20 μm. The distribution of the charges on the strips allows a high spatial resolution in one coordinate, of the order of a few μm. The spatial resolution of the second coordinate is achieved by using orthogonal microstrips on the other side. The readout of microstrip detectors is performed using dedicated electronics bonded to the detector, that include preamplification. Semiconductor microstrip detectors offer several advantages: they are compact detectors, with an high stopping power; they offer high spatial resolution, of the order of a few μm.

7.2   Charge Coupled Devices (CCD)

The *Charge Coupled Devices* (CCD) have been developed for the imaging of faint levels of optical radiation. They have quickly become the standard instrument of optical astronomy. The incident photon produces electron-hole pairs. The electrons are collected by a bias electric field and stored inside the pixels. The collected charge is shifted from pixel to pixel to an amplifier and an analog to digital converter. The CCDs are sensitive to X-rays [1, 3, 4, 8] and have been used in several X-ray space based observatories. The standard CCDs based on CMOS technology are made more sensitive to X-rays increasing the thickness of the depletion layer. The CCDs can be used with front illumination or back illumination. The lower energy X-rays are blocked by the layer with the gates. A possible solution is the back illumination, with the thinning of the detector to achieve full depletion. The other solution is the thinning of the gate layer or the etching of holes in the oxide layer (open gate structure). The back illuminated CCDs show a larger quantum efficiency compared to front illuminated CCDs. The X-ray CCDs are equipped with filters to block optical light, that is a source of noise. The CCDs used in X-ray astronomy perform charge transfer through frame transfer. The incident X-rays produce photons in the image section during the frame time, the equivalent of the exposure time in optical astronomy. The stored charges are transferred row by row to the frame store section, isolated and shielded from the X-ray flux. The content of the rows of the frame store are transferred to a readout register, followed by the amplifier and digitizer system.

The CCDs have found a widespread application in the focal plane of grazing incidence telescopes in space (Chap. 11). Due to the typical dimensions of the detector, some square cm^2, it is necessary to build arrays of several CCDs in the focal plane for imaging and for spectroscopy. The CCDs are operated in photon counting mode, recording the position and the energy of the incident X-ray photons, differently from the optical domain, where the signals of the photons are accumulated during long exposures. The minimum frame time of CCDs for X-rays can be very short, of the order of a few seconds, thus most pixels show low count rates. The raw CCD frames are processed on board. Firstly the bias level, the signal in absence of external X-ray flux, is removed. Then the content of each pixel is compared to a predefined threshold: if the pulse amplitude exceeds the threshold, the event is accepted. The properties of the event, i.e. position, occurrence time, pulse height (a proxy for energy) are recorded, together with the pulse heights of the near pixels. The morphology of the pixel involved in the event allows the discrimination between signals induced by genuine X-rays and by the background of charged particles. The events induced by the former are more localized than those produced by cosmic rays, since an X-ray is promptly absorbed, while a charged particle produces an ionization signal along the whole path in the detector, that undergoes diffusion. A cut based on the event morphology reduces the background.

Problems

7.1 Discuss the different strategies for the detection of high energy photons and charged particles with semiconductor detectors.

7.2 Estimate the number of electron-hole pairs produced by a minimum ionizing particles in 300 μm of silicon and germanium.

References

1. Grupen, C. and Swartz, B.: Particle Detectors. Cambridge University Press (2008)
2. Grupen, C. and Buvat, I: Handbook of Particle Detection and Imaging. Springer-Verlag Berlin Heidelberg (2012)
3. Huber, M. C. E., Pauluhn, A., Culhane, J. L., Gethyn T. J., Wilhelm, K., Zehnder, A.: Observing Photons in Space - A Guide to Experimental Space Astronomy. Springer Science+Business Media, New York (2013)
4. Lèna, P. et al.: Observational Astrophysics. Springer-Verlag Berlin Heidelberg (2012)
5. Leo, W. R.: Techniques for Nuclear and Particle Physics Experiments - A How-to Approach. Springer-Verlag (1994)
6. Olive, K.A. et al. (Particle Data Group): Chin. Phys. C **38**, 090001 (2014)
7. Schoönfelder, V.: The Universe in Gamma Rays. Springer-Verlag Berlin Heidelberg (2001)
8. Trumpër, J.E. and Hasinger, G.: The Universe in X-Rays, Springer-Verlag Berlin Heidelberg (2008)

Chapter 8
Cherenkov and Transition Radiation Detectors

Cherenkov detectors are based on the radiation emitted when a charged particle is moving through a medium faster than the speed of light in the medium itself. The emitted photons, created in a very short time and emitted along the particle velocity, are detected with photon transducers. The Cherenkov detectors measure the velocity of particle. Transition Radiation Detectors (TRD) are based on the emission of radiation by highly relativistic charges that traverse media with different dielectric constants, as a set of packed foils. The TRDs provide a measure the relativistic gamma factor of the particle. Both classes of detectors are used for particle identification.

8.1   Cherenkov Detectors

The simplest Cherenkov detector is a threshold detector, whose threshold is determined by the refractive index of the radiator [1–3]. The measurement of the Cherenkov angle θ_c provides the measurement of the particle velocity. The *Ring Imaging Cherenkov* (RICH) detector (Fig. 8.1) uses a spherical mirror with radius R_m to focus the Cherenkov light, producing a ring that is sensed with a spherical detector with radius $R_d = \frac{R_m}{2}$. The volume separating the two spherical surfaces is filled with the radiator material, whose refractive index selects the momentum interval of particles that are identified.

The radius of the radiation ring is:

$$r_{ring} = \tan\theta_c \frac{R_m}{2} \tag{8.1}$$

The particle velocity is estimated as:

$$\beta = \frac{1}{n}\frac{1}{\cos\left(\frac{2r_{ring}}{R_m}\right)} \tag{8.2}$$

© Springer International Publishing Switzerland 2017
R. Poggiani, *High Energy Astrophysical Techniques*,
UNITEXT for Physics, DOI 10.1007/978-3-319-44729-2_8

Fig. 8.1 Layout of a RICH
detector

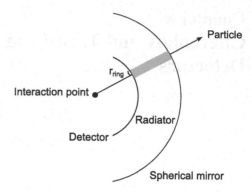

The detector used to map the radiation ring must have a good spatial resolution. Multi Wire proportional chambers are a popular choice, with the addition of photosensitive materials like TMAE ($[(CH_3)_2N]_2C$) and TEA ($(C_2H_5)_3N$) to the gas.

A different kind of Cherenkov detector widely used in high energy astrophysics is the water Cherenkov detector, that measures the Cherenkov light emitted by charged particles in water, whose amount is proportional to the path length. The Cherenkov radiation is measured with photomultiplier tubes. The Cherenkov threshold in water is $\beta = 0.75$. Water must very pure and must be protected against contamination. The water Cherenkov detector is used for the detection of electrons and muons produced by the interactions of neutrinos (Chap. 14).

8.2   Transition Radiation Detectors

As discussed in Chap. 2, the transition radiation emitted at the crossing of a single interface is very small. A Transition Radiation Detector (TRD) is built by stacking a large number of thin foils to have a large number of vacuum to matter transitions (Fig. 8.2) [1–3].

Fig. 8.2 Transition
Radiation Detector

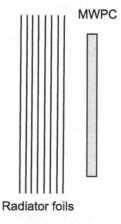

The foil material must have a low atomic number to avoid the photoelectric absorption of the emitted photons inside the detector itself. The stacks are followed by a detector for X-rays, for example a proportional chamber filled with a gas with high atomic number. The thickness of the foils cannot be arbitrarily reduced, since it must exceed the formation length for radiation, of the order of $\frac{\gamma c}{\omega_p}$. The TRD are very popular for the particle identification in the direct measurements of cosmic rays in observatories outside the atmosphere (Chap. 13).

Problems

8.1 Discuss the difference between detectors based on Cherenkov radiation and Transition Radiation.

References

1. Grupen, C. and Swartz, B.: Particle Detectors. Cambridge University Press (2008)
2. Grupen, C. and Buvat, I: Handbook of Particle Detection and Imaging. Springer-Verlag Berlin Heidelberg (2012)
3. Olive, K.A. et al. (Particle Data Group): Chin. Phys. C **38**, 090001 (2014)

Chapter 9
Calorimeters

This chapter deals with the calorimetric techniques for the measurement of the energy of particles, relying on their absorption. High energy photons and nuclei produce showers of secondary particles in a material (Fig. 9.1). The energy resolution of the various classes of calorimeters is discussed. The processes are used to build electromagnetic and hadron calorimeters, respectively. An ideal calorimeter should contain the whole shower. Calorimeters can be homogeneous, made of materials acting as absorbers and detectors at the same time, or sampling, with thin detectors separated by absorber layers. The energy resolution of the various classes of calorimeters is discussed. The energy of very low energy particles is measured with cryogenic calorimeters, that measure the rise in temperature produced by the absorption of the particle.

9.1 Electromagnetic Calorimeters

The *electromagnetic calorimeters* [1–3] are based on the mechanisms of photon interactions above the MeV energy range, the pair production and the bremsstrahlung discussed in Chap. 2. The close value of the radiation length X_0 and the pair production length λ_{pair} is the responsible of the development of the electromagnetic showers. The complete containment of an electromagnetic shower requires a thickness of the order of twenty radiation lengths and a transverse dimension of a few Moliere radii.

The longitudinal profile of the energy deposition in a calorimeter is given by [1–3]:

$$\frac{dE}{dt} = E_0 \frac{b(bt)^{a-1}e^{-bt}}{\Gamma(a)} \tag{9.1}$$

where E_0 is the energy of the primary photon, Γ is the Euler function and the depth of maximum is $\frac{a-1}{b}$.

© Springer International Publishing Switzerland 2017
R. Poggiani, *High Energy Astrophysical Techniques*,
UNITEXT for Physics, DOI 10.1007/978-3-319-44729-2_9

Fig. 9.1 Shower
development in a calorimeter

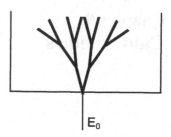

The energy resolution of a calorimeter depends on the energy, according to:

$$\frac{\sigma(E)}{E} = \frac{A}{\sqrt{E}} \oplus \frac{B}{E} \oplus C \qquad (9.2)$$

where $\oplus$ defines the addition in quadrature. The first contribution is due to the fluctuations of the photoelectrons produced in the absorption process; the second contribution is produced by the noise of the electronics; the third contribution is caused by the uncertainty in the calibration of the calorimeter and by non uniformities. If the first term is dominant, the energy resolution improves with energy as $\frac{1}{\sqrt{E}}$.

The *homogeneous calorimeters* are built with a material that acts as the absorber and as the detecting medium at the same time, providing sensitivity over the whole volume. The measure of energy by the calorimeter relies on different processes: ionization, scintillation, emission of Cherenkov radiation. The *crystal calorimeters* are based on scintillator materials with high density, such as PbWO$_4$, coupled to photomultiplier tubes. Crystal calorimeters achieve the best resolution among calorimeters, where A is of the order of a few percent. On the other hand, crystal calorimeters require large quantities of expensive materials that should be highly homogeneous. Homogeneous calorimeters relying on ionization detector arrays inside liquefied noble gases, such as Kripton and Xenon, achieve a comparable resolution.

The high cost of homogeneous calorimeters has prompted the development of *sampling calorimeters*, consisting of arrays of thin detectors interleaved with layers of absorbing materials. The energy released by the event is sampled: the process of sampling introduces an additional source of fluctuations. The number of crossings of the detector layers is:

$$N_{tot} = \frac{\tau}{d_d} \qquad (9.3)$$

where τ is the total track length and d_d the thickness of a layer consisting of the absorber and the detector. The effectively measurable track length is:

$$\tau_m = X_0 \frac{E}{E_c} F_\xi \qquad (9.4)$$

where the function F_ξ takes into account the cut-off on the measurable track length. The number of track segments is:

$$N = F_\xi \frac{E}{E_c} \frac{X_0}{d_d} \tag{9.5}$$

The sampling fluctuations set a limit on the energy resolution, given by:

$$\left(\frac{\sigma(E)}{E}\right)_{sampling} = \sqrt{\frac{E_c \, d_d}{E \, X_0 \, F_\xi}} \tag{9.6}$$

The resolution is proportional to $E^{-\frac{1}{2}}$ as in homogeneous calorimeters. The resolution of sampling calorimeters is worse than that of homogeneous calorimeters, with a coefficient of the order of 10–20 %.

9.2 Hadron Calorimeters

The *hadron calorimeters* rely on the development of an hadronic shower in the material [1–3]. The evolution of hadron showers is governed by the nuclear interaction length λ_{int}, that is generally larger than the radiation length. The longitudinal development of the hadronic cascade is longer than the longitudinal development of the electromagnetic shower. The hadron calorimeters are thicker than electromagnetic calorimeters. The lateral development of hadronic showers is larger than that of an electromagnetic calorimeter, since there are large transverse momentum transfers, of the order of 0.35 GeV/c. The hadronic shower in the calorimeter contains a large variety of particles: nuclear fragments, pions, muons, neutrinos, photons, electrons/positrons, etc. One third of the pions produced at each interaction are neutral pions that decay into photons and start electromagnetic subshowers. After m interactions, there will be an electromagnetic fraction:

$$f_{em} = 1 - \left(\frac{2}{3}\right)^m \tag{9.7}$$

The production of neutral pions shows large fluctuations. Another component of the shower is the fraction f_h of charged hadrons, that can be detected by their ionization energy loss. In addition to the discussed fractions that can be detected, there is a fraction f_{inv} of *invisible energy* that is used to break nuclear bonds and is related to neutrinos. The fraction of invisible energy can be as high as 40 %. The large fluctuations in the development of the hadronic shower produce a worse energy resolution for hadronic calorimeters. The problem of invisible energy can be mitigated by using the technique of *compensation*. If the absorber material is uranium or a similar material, there will be production of neutrons that will produce further neutrons and high energy photons, injecting additional energy. Generally, hadron calorimeters are sampling calorimeters with active layers using the same technology of the detector layers in the electromagnetic calorimeters: scintillators, cryogenic liquids, gas chambers. The best resolution achieved by hadron sampling calorimeters is of the order of:

$$\left(\frac{\sigma(E)}{E} \right)_{hadron} = \frac{40\,\%}{\sqrt{E(GeV)}} \tag{9.8}$$

The energy resolution improves with energy as $\frac{1}{\sqrt{E}}$.

9.3 Cryogenic Calorimeters

The calorimeters described above are used for high energy particles. The cryogenic microcalorimeters are used in the X-ray region and show intrinsic energy resolution. The spectroscopy of particles in the low energy region, below 1 MeV, covers a wide range of astrophysical topics of interest, from the X-ray spectroscopy to the search for dark matter candidates. Microcalorimeters for low energy particles operate at cryogenic temperatures, in the milliKelvin range. The layout of a microcalorimeters is shown in Fig. 9.2.

The deposition of an energy amount ΔE in the absorber produces a temperature increase:

$$\Delta T = \frac{\Delta E}{C} \tag{9.9}$$

where C is the heat capacity of the absorber. The temperature increase is measured with a thermometer integrated with the absorber. The detector is connected to a thermal bath through a low thermal conductivity link. The limit to the resolving power is given by:

$$\Delta E \sim \sqrt{\frac{k_B T_t C}{\alpha}} \tag{9.10}$$

where T_t is the temperature of the thermal bath and α is the responsivity of the thermometer. The spectral resolution improves choosing detectors with low thermal capacity and operating at low temperature. Semiconductor thermistors allow to achieve spectral resolutions better than one percent. They have been used in the

Fig. 9.2 Layout of a cryogenic calorimeter

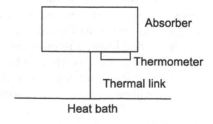

Suzaku Observatory, with a resolution of a few eV at 6 keV. The *Transition Edge Sensors* (TES) thermistors are based on a superconducting film operated close to the transition temperature: a small variation in the temperature produces a large variation of the resistance at the transition from superconductor to metal. The spectral resolution is very high, of the order of 1 or 2 eV at 6 keV.

Problems

9.1 Discuss the main differences between the electromagnetic and the hadron calorimeters.

9.2 Discuss the factors limiting the energy resolution in electromagnetic and the hadron calorimeters.

References

1. Grupen, C. and Swartz, B.: Particle Detectors. Cambridge University Press (2008)
2. Leo, W. R.: Techniques for Nuclear and Particle Physics Experiments - A How-to Approach. Springer-Verlag (1994)
3. Olive, K.A. et al. (Particle Data Group): Chin. Phys. C **38**, 090001 (2014)

Chapter 10
Measurement of Physical Properties of Photons and Particles

The previous chapters have presented the photon and particle detectors from the point of view of the radiation-matter interactions. This chapter revisits the detectors from the point of view of the physical quantities of interest for high energy astrophysics: position, time, energy, momentum, particle identification. The chapter presents the detectors used for position measurement, or tracking detectors, based on ionization in gas, liquids or solid state materials. Time of Flight (TOF) can be measured by using pairs of detectors with fast response time. The energy of a particle is measured using calorimeters. The momentum of charged particles is measured tracking their motion inside a magnetic field. Particle identification can be performed coupling the measurement of the momentum with the measurement of energy loss by ionization, of the velocity or of the total energy. The measurement of some observables requires the combination of detectors. The new approach will be used in later chapters to discuss the general purpose instrumentation used for ground and space based observatories.

10.1 Position and Tracking

The reconstruction of particle trajectories is of paramount importance for high energy astrophysics, to investigate the position of the interaction of interest and to measure the path of produced particles. The spatial resolution of some typical detectors is reported in Table 10.1. The gas detectors and the semiconductor based detectors are the most used solutions [1–4].

© Springer International Publishing Switzerland 2017
R. Poggiani, *High Energy Astrophysical Techniques*,
UNITEXT for Physics, DOI 10.1007/978-3-319-44729-2_10

Table 10.1 Spatial resolution of particle detectors; data from [4]

Detector	Spatial resolution (μm)
Proportional chamber	50–100
Drift chamber	100
Scintillator	100
Silicon strip	A few

Table 10.2 Time resolution of particle detectors; data from [4]

Detector	Time resolution
Proportional chamber	2 ns
Drift chamber	2 ns
Scintillator	100 ps
Silicon strip	Some ns

10.2   Time of Flight

The measurement of the *Time of Flight* (TOF) between two positions at distance D is used to measure the velocity of a particle and to provide the identification of particle with known momenta and different masses [1–4]. Two detectors at the two positions provide the start and the stop. Two particles sharing the same momentum p, but having different masses M_1, M_2, show a difference in the time flight by a quantity:

$$\Delta t = \frac{D}{c} \left(\sqrt{1 + \frac{m_1^2 c^2}{p^2}} - \sqrt{1 + \frac{m_2^2 c^2}{p^2}} \right) \qquad (10.1)$$

The time resolution of some detectors is reported in Table 10.2. The most used detectors for TOF systems are plastic scintillators read by photomultiplier tubes.

10.3   Energy

The calorimeters [1–4] show two advantages for the measurement of the energy. Their longitudinal size increase only logarithmically with energy, thus they are compact instruments. In addition, the energy resolution $\frac{\sigma(E)}{E}$ behaves as $\frac{1}{\sqrt{E}}$, improving with increasing energy. The resolution of some electromagnetic calorimeters used in particle and astroparticle physics is shown in Table 10.3.

Detector	Time resolution
NaI(Tl) (Crystal Ball)	$\frac{2.7\%}{E^{\frac{1}{4}}}$
PbWO$_4$ (CMS)	$\frac{3\%}{\sqrt{E}} \oplus 0.5\% \oplus \frac{0.2}{E}$
LAr/Pb (ATLAS)	$\frac{10\%}{E} \oplus 0.4\% \oplus \frac{0.3}{E}$
AMS-02 Ecal	$\frac{0.104}{\sqrt{E}} \oplus 0.014$

Table 10.3 Energy resolution of some electromagnetic calorimeters, with energy in GeV; data from [4]

10.4 Momentum

The momentum of a charged particle is measured using *magnetic spectrometers* [1, 4]. The magnetic field deviates the particle along an helix trajectory whose axis is directed along the magnetic field B. The sign of the particle charge is determined by the direction of the helix. Tracking detectors are installed inside the magnetic field to sample the particle position at N points along a track of total length L with a spatial resolution σ. The magnetic field B acts only on the transverse component of the momentum. The error due to the track measurement at N points is [1]:

$$\left(\frac{\sigma(p)}{p}\right)_{tracking} = \frac{\sigma[m]}{0.3B[T]L[m]^2}\sqrt{\frac{720}{N+4}}p[GeV/c] \tag{10.2}$$

The error due to the multiple scattering [1]:

$$\left(\frac{\sigma(p)}{p}\right)_{scattering} = \frac{13.6\sqrt{L/X_0}}{e \int B\,dl}MeV/c \tag{10.3}$$

where X_0 is the radiation length of the material. The momentum resolution is proportional to the magnetic field and to the square of the track length and becomes worse with increasing momentum, thus magnetic spectrometers are less performant than calorimeters at high energies.

10.5 Particle Identification

The identity of a particle requires the measurement of its charge and its mass, in addition to the measurement of the energy or the momentum. The particle identification requires a combination of detectors. As discussed above, the measurement of the time of flight allows the discrimination between particles with identical momentum and different masses. The mechanisms of energy loss of charged particles discussed in Chap. 2 depend on the charge z and on the energy: ionization, Cherenkov radiation, Transition Radiation. Generally the identification of the particle type by the ionization energy loss is possible below the ionization minimum. The Cherenkov detectors are a common choice for particle identification. The RICH detectors discussed in Chap. 8 allow the determination of the particle velocity by the size of the Cherenkov ring. If the momentum is known, then the particle mass can be estimated. The TRDs produce a signal proportional to the energy of the particle, but have an high threshold. The calorimeters provide particle identification through the different evolution of the longitudinal and lateral development of the electromagnetic and hadronic showers. Calorimeters with the capability to measure single particles and to estimate their energy can discriminate between charged and neutral particles by the presence or absence of a track in a tracking system. Electromagnetic calorimeters can separate electrons and photons looking for the presence of lack of a track in an external tracking system.

Problems

10.1 Discuss the possible detector solutions for the tracking of charged particles.

10.2 Discuss the measurements of energy and momentum.

References

1. Grupen, C. and Swartz, B.: Particle Detectors. Cambridge University Press (2008)
2. Grupen, C. and Buvat, I: Handbook of Particle Detection and Imaging. Springer-Verlag Berlin Heidelberg (2012)
3. Huber, M. C. E., Pauluhn, A., Culhane, J. L., Gethyn T. J., Wilhelm, K., Zehnder, A.: Observing Photons in Space - A Guide to Experimental Space Astronomy. Springer Science+Business Media, New York (2013)
4. Olive, K.A. et al. (Particle Data Group): Chin. Phys. C **38**, 090001 (2014)

Part III
High Energy Astronomy

Chapter 11
Ultraviolet and X-Ray Astronomy

This chapter discusses the astrophysics with photons having an energy up to hundreds of keV. X-rays are not able to penetrate atmosphere and space based observatories are needed. The ultraviolet part of the spectrum is an hybrid, from the point of view of instrumentation, of the high and low energy astronomy techniques. UV detectors are generally based on photoconduction. On the other hand, telescopes are very similar to telescopes for soft X-rays. Telescopes for UV and soft X-rays use reflective elements based on grazing incidence. Higher energy X-rays are collected using collimator telescopes or coded aperture masks. The detectors used for X-rays are: scintillators, gas detectors and solid state detectors, including CCDs. X-ray spectroscopy is performed using the intrinsic energy resolution capability of detectors or diffraction gratings as in optical spectroscopy. X-ray polarimetry is also discussed. The main UV and X-ray observatories will be discussed as case studies.

11.1 Typical Fluxes

The X-ray emission of astrophysical sources is produced by thermal and non thermal processes. Several classes of objects involve plasmas optically thin to their own radiation and at high temperature, from 10^6 to 10^8 K, that produce *thermal emission*. The continuum emission spectra are produce by bremsstrahlung and radiative recombination. The spectra show emission lines produced by the a variety of elements in different ionization states. The *non thermal emission* is produced by electrons with a power law distribution in energy that undergo non thermal bremsstrahlung in dense plasmas [6].

The X-ray flux densities of some sources, measured in milliCrab units with reference to the Crab flux, are reported in Table 11.1.

© Springer International Publishing Switzerland 2017
R. Poggiani, *High Energy Astrophysical Techniques*,
UNITEXT for Physics, DOI 10.1007/978-3-319-44729-2_11

Table 11.1 Flux density of some X-ray sources [1]

Source	Flux density (millicrab)	Energy range (keV)	Source type
Crab	1000	0.2–4	SNR
Sirius	0.34	0.2–4	Star
Cassiopeia A	89	0.2–4	SNR
3A 0620-00	1,250,000	2–10	X-ray binary
Sco X-1	17,000	2–10	X-ray binary
Cyg X-1	1175	2–10	X-ray binary
M31	2.3	0.2–4	Normal galaxy
3C 273	3.1	2–10	Active galaxy
M87 (Virgo)	22	2–10	Cluster of galaxies

11.2 Telescopes for Hard X-Rays

The UV and X-rays are absorbed in matter, thus it is not possible to use the traditional techniques of refracting and reflecting optics familiar to optical astronomy to build X-ray telescopes [2, 3, 5, 6]. Only ultraviolet and low energy X-rays, below a few keV, can be reflected with grazing incidence. Telescopes for higher energy X-rays are collimator like, with the X-ray detector placed at the bottom: the field of view is set by the geometrical aperture of the collimator. The imaging of the field of view is achieved by tuning the blockage or the transmission of the incident radiation according to its direction. The angular distribution is mapped into a spatial distribution or a time variation of the photon flux at the detector. The masking systems can use a single absorbing unit with some apertures or different layers of grids.

The *spatial aperture modulation* uses a pattern of randomly positioned opaque and transparent areas in the aperture. The instruments are called *coded mask telescopes*, since the pattern is a mask designed with a specific code of transmitting and blocking regions (Fig. 11.1). A position sensitive detector, that must have a spatial resolution better than the size of the apertures, measures the shadow of the mask. The effective area of the telescope is of the order of the transparent area of the mask. The Fully Coded Field of View (FCFOV) system consists of a mask and a detector with equal sizes: only photons that have crossed the mask arrive at the detector. The Partially Coded Field of View (PCFOV) system uses a mask larger than the detector, to increase the observable field of view. The pattern of the aperture is random; the most used one is the *Uniformly Redundant Array* (URA), that has been used in the INTEGRAL and Swift Observatories.

The spatial resolution of a coded mask instrument is of the order of the ratio of the size of the apertures to the separation between mask and detector. The imaging of the sky field is achieved by the inversion of the *coding equation*:

$$O(\mathbf{r}) = M(\mathbf{r}) * I(\mathbf{r}) \qquad (11.1)$$

Fig. 11.1 An example of coded mask

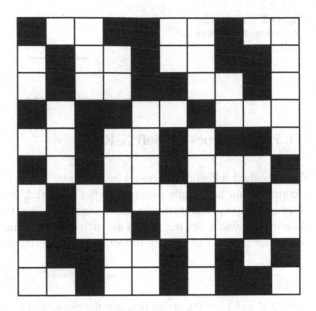

where **r** is the vector describing the two dimensional position, O is the observed intensity distribution at the detector, I is the intrinsic brightness distribution, M the aperture modulation function. The inversion of the coding equation (Eq. 11.1) is performed by the *cross correlation* of the aperture modulation function with the observed intensity distribution. In the *back-projection* method the mask pattern is projected into the sky to assess the probable original position of each photon. The successive application of the algorithm provides additional clues to the origin of each photon, thus the image of all sources in the field.

The *Modulation collimator* replaces the mask with two or more grids, with rods or slats opaque to X-rays alternated to clear slits (Fig. 11.2). The imaging is achieved by scanning the optical axis orthogonally to the slat direction, thus modulating the intensity of the sources in the field of view. The grid collimators with two orthogonal planes of parallel slats mounted parallel to the detector allow the reconstruction of two coordinates by performing two separate scans along two directions. In the temporal aperture modulation technique the motion of the aperture produce a variation of the signal at the detector, that does not need to be position sensitive. The *rotation modulated collimator* (RMC) uses a double grid rotating around the optical axis, that produces a modulation in time. Each X-ray source in the field of view will produce a distinct modulation curve. The image is reconstructed by assuming that the field of view contains only point like sources and computing the cross correlation function of the observed curve with the response function of the collimator.

Fig. 11.2 Layout of an
aperture modulator with two
grids

11.3   Telescopes for Soft X-Rays

The standard reflection processes instrumental in the construction of optical telescopes cannot be directly extended to the collection of UV and soft X-rays, since
they are able to penetrate the candidate materials where they are absorbed. Reflection
is only possible when the X-rays are arriving to the material surface with *grazing
incidence* [3, 6]. The index of refraction of a material is:

$$n = 1 - \delta - i\beta \tag{11.2}$$

where β and δ are the absorption and the phase change, that depend on the atomic
scattering factors f_1, f_2 [3]. The quantity δ is approximated by [3]:

$$\delta = \frac{1}{2\pi} r_e \frac{N_A \rho}{A} Z \lambda^2 \tag{11.3}$$

where ρ, A are the density and the atomic mass of the material, N_A is the Avogadro
number, r_e is the classical electron radius. The angles are conventionally measured
relative to the plane of the surface and not relative to the normal, defining the grazing
angle as the complement to $\frac{\pi}{2}$ of the standard angles of the Snell reflection law. There
will be total reflection for angles smaller than the *critical grazing angle* ϕ_c, where:

$$\cos \phi_c = 1 - \delta \tag{11.4}$$

The critical angle is approximated by $\phi_c \sim \sqrt{2\delta}$. The critical grazing angle is
inversely proportional to the energy of the incident X-ray, making reflection more
and more difficult above a few keV. The critical angle is also proportional to $\sqrt{Z}$,
where Z is the atomic number of the material: high Z materials allow the reflection
of X-rays with higher energies at a fixed grazing angle. The reflectivity of gold for
grazing angles of 0.5° and 1° is reported in Fig. 11.3 for reference.

The reflecting surfaces for X-ray astronomy are the combination of a substrate,
a material with a low coefficient of thermal expansion, and of a single or multilayer
coating. X-rays incident at angle larger than the grazing angle have a small probability
for reflection at a single layer, but the combination of hundreds layers can build
up an appreciable reflectivity over a limited wavelength interval. The multi-layer
mirrors use alternate layers of high Z materials, that have a high reflectivity, and
of low Z material that have a small absorption. Despite the narrow band width, the

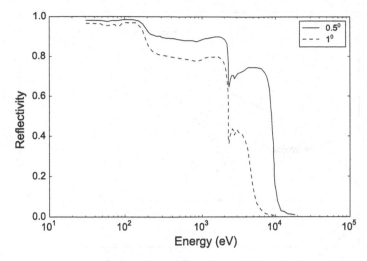

Fig. 11.3 Reflectivity of gold for grazing angles of 0.5° and 1°; data from http://xdb.lbl.gov/

configuration can be used up to some tens keV. The incident X-rays can undergo scattering due to the rugosity of the surface, producing an halo around the sources. The scattering is proportional to the square of the energy of the X-ray. The scattering intensity, relative to the total power at the focal plane, is of the order of $(\frac{4\pi\sigma\sin\phi}{\lambda})^2$, where σ is the rms roughness of the reflecting surface. The contamination of the optical surfaces by the presence of dust reduces the performances of the reflecting surface and must be kept as low as possible.

The parabola provides a natural choice for the substrate shape, since photons incident along its axis will be reflected towards the focus. If ϕ is the grazing angle, then the X-ray will be deflected by an angle 2ϕ. The parabolic surface shaped as a plate has been used in the concept of the *Kirkpatrick-Baez* system. Two parabolic plates are mounted at right angles, providing focusing in two directions. The small effective collecting area has prevented a wide use in X-ray astronomy. Telescopes for soft X-rays use two mirrors. A paraboloid mirror is coupled with a confocal hyperboloid, in the *Wolter type I telescope* (Fig. 11.4). The X-rays are firstly reflected by the paraboloid and then reflected to the main focus by the hyperboloid.

The collection area of a Wolter type I telescope is limited by the mechanism of grazing incidence to an annulus close to the conic surfaces. The area is increased by using nested paraboloid-hyperboloid shells. The requirement on the quality of optical surfaces are tight, due to the small wavelength of the X-rays. The mirrors of the Wolter telescopes are manufactured with different techniques. The most precise surfaces are obtained by machining, polishing and coating glass blanks, as in Chandra. Alternative and cheaper solutions are the coating of electroformed metal shells replicated from templates, as in XMM-Newton, or the assembling of foil mirror segments, as in ASCA.

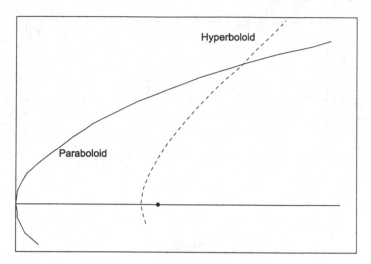

Fig. 11.4 Layout of a Wolter type I telescope

11.4 UV and X-Ray Spectroscopy

The spectroscopy in the X-ray domain is often performed by the same detectors that provide the imaging of the sky field, when they have intrinsic resolution capability, as proportional counters and CCDs. The direct spectroscopy mode is a suitable choice for the harder X-rays, since the energy resolution improves with increasing energy. Microcalorimeters operating below 100 mK can measure X-ray photons and have been adopted for the Suzaku spectrometer.

The spectroscopic observations in the region of UV and soft X-rays can be secured by using the same technique of optical astronomy, i.e. dispersing elements (see [4], for example). The most used element is the *diffraction grating*, either in transmission or in reflection. The grating equation is:

$$\sin \alpha + \sin \theta = \frac{m\lambda}{\sigma} \tag{11.5}$$

where α is the incidence angle, θ the diffraction angle, σ the grating spacing, m the diffraction order. Usually gratings have hundreds or thousands lines per millimeter and operate in the first order of diffraction. The diffraction gratings have an almost constant wavelength resolution $\Delta\lambda$, thus the resolving power improves with increasing wavelength, i.e. with decreasing energy.

11.5 X-Ray Polarimetry

The polarization of the X-rays is a probe of different processes, among them the nature of sources of strong gravity [3]. The *X-ray polarimetry* in the region up to a few tens keV uses the Bragg reflection or the photoelectric effect. The Bragg reflection has a maximum for photons with an electric field parallel to the crystal planes; the Bragg condition limits the use of the method to a narrow range in energy. The angular distribution of the electrons extracted in the photolectric effect is a function of the cosine of the angle χ between its direction and the direction of the electric field of the photon:

$$\frac{d\sigma}{d\Omega} \sim \frac{\sin^2\theta \cos^2\chi}{(1 - \beta \cos\theta)^4} \tag{11.6}$$

where θ is the polar angle of electron. The polarization of photons with energies larger than some tens keV is measured with the Compton scattering; the angular distribution of the scattered photon depends on the direction of its polarization:

$$\frac{d\sigma}{d\Omega} \propto \left(\frac{E_f}{E_i}\right)^2 \left(\frac{E_i}{E_f} + \frac{E_f}{E_i} - 2\sin^2\theta \cos^2\zeta\right) \tag{11.7}$$

where E_i, E_f are the initial and final energy of the photon, θ is the scattering angle and ζ is the angle between the direction of the photon polarization and the plane of scattering.

11.6 UV and X-Ray Observatories

Some observational facilities in the UV and X-rays are presented here as example. Some space observatories complement the X-ray instrumentation with optical instrumentation on board, to perform multiwavelength observations and the optical follow-up of transient events, such as the GRBs.

The *RXTE* mission[1] used a combination of detectors. The HEXTE detector was a double cluster with four phoswich detectors (NaI(Tl) with CsI(Na)) each, for the energy range 15–250 keV, achieving an energy resolution of 15 % at 60 keV. The PCA detector was an array of proportional counters with a collection area of $6500\,\text{cm}^2$ and an energy resolution better than 18 % at 6 keV. The ASM consisted of three wide angle shadow cameras coupled to position sensitive proportional counters for the energy range 2–10 keV.

The *INTEGRAL* Observatory[2] is a coded mask instrument. The SPI Spectrometer uses cooled Ge detectors. IBIS is an imager with two layers of detector arrays, the

---

[1] https://heasarc.gsfc.nasa.gov/docs/xte/XTE.html.
[2] http://sci.esa.int/integral/.

CdTe one with 16,000 pixels and the CsI one with 4000 pixels. The two instruments have polarization capability. JEM-X is an X-ray monitoring instrument with a position sensitive microstrip proportional counter for the energy interval 3–35 keV. The OMC is a CCD optical monitor with CCD and lens optics. The four instruments observe the same region of the sky simultaneously.

The *Chandra Observatory*[3] uses a Wolter type I telescope with four pairs of nested mirrors. Two instruments are in the focal plane, ACIS and HRC. The Chandra Advanced CCD Imaging Spectrometer (ACIS) is an array of CCDs that provides at the same time the imaging and the energy measurement of the X-rays. The High Resolution Camera (HRC) consists of two Micro-Channel Plates and has a resolution of half an arc second. The two imaging systems are completed by two spectrometers, the High Energy Transmission Grating Spectrometer (HETGS) and the Low Energy Transmission Grating Spectrometer (LETGS). The diffracted X-rays are measured with ACIS or HRC. The period of the LETG and HETG gratings 1 and 0.2 μm, respectively.

The *XMM-Newton* Observatory[4] consists of three separate X-ray telescopes. At the prime focus of each of them there is a European Photon Imaging Camera (EPIC), a silicon CCD with fast timing capability. Two of the three telescopes have a Reflection Grating Spectrometer (RGS) system for spectroscopy. The observatory is equipped with an optical/UV telescope (aperture 30 cm), the Optical Monitor (OM), to observe the same area of the sky as the X-ray telescopes, but at ultraviolet and visible wavelengths, to produce complementary data about the X-ray sources. The telescope is mounted on the mirror support platform of XMM-Newton.

The *Swift* observatory[5] is dedicated to the study of Gamma Ray Burst with a suite of three instruments. The Burst Alert Telescope (BAT) is a coded aperture imaging system equipped with CdZnTe detectors that cover the region from 15 to 150 keV with a field of view of 2 sr and an angular resolution of the order of four arc minutes. The X-ray Telescope (XRT) is a Wolter I instrument equipped with a CCD for imaging and spectroscopy in the low energy region from 0.3 to 10 keV. The UV/Optical Telescope (UVOT) secures UV and optical imaging and spectroscopy of GRB afterglows, with a CCD and grism combination.

The *MAXI* observatory[6] is a mission on the International Space Station (ISS). MAXI uses two types of X-ray imaging systems, the Gas Slit Camera (GSC) and the Solid-state Slit Camera (SSC). Both cameras use slit and collimator optics. The GSC systems are made of six identical units; each unit comprises a slit and slat collimator and two proportional counters. The SSC is a system consisting of two CCD cameras.

NuSTAR[7] is a Wolter I telescope for the energy range from 3 to 79 keV. NuSTAR has two detector systems that comprise four CdZnTe with an energy resolution of 0.4

---

[3] http://chandra.harvard.edu/.

[4] http://sci.esa.int/xmm-newton/.

[5] http://swift.gsfc.nasa.gov/.

[6] http://maxi.riken.jp/top/.

[7] http://www.nustar.caltech.edu/.

keV. The detectors are shielded by CsI scintillators for rejection of the background of photons and X-rays.

The future mission *eROSITA*[8] will perform the first imaging all-sky survey in the energy region up to 10 keV. It will use seven identical Wolter 1 mirror modules and the frame pnCCDs.

The telescopes for the Extreme UltraViolet (EUV) region are Wolter type I telescopes, but with larger grazing angles, of the order of a few degrees. The region was initially believed to be not accessible due to the opacity of the interstellar medium, but there are inhomogenities that leave some regions for observations. Normal incidence telescopes can be used only above 500 Å. Some notable missions are the *International Ultraviolet Telescope* (IUE),[9] *Far Ultraviolet Spectroscopic Explorer* (FUSE), *Galaxy Evolution Explorer* (GALEX), *Extreme Ultraviolet Explorer* (EUVE).

Problems

11.1 Discuss the Wolter type I telescope.

11.2 Discuss the differences between the telescopes for soft X-rays and hard X-rays.

References

1. Cox, A. N.: Allens Astrophysical Quantities. Springer (2002)
2. Grupen, C. and Buvat, I: Handbook of Particle Detection and Imaging. Springer-Verlag Berlin Heidelberg (2012)
3. Huber, M. C. E., Pauluhn, A., Culhane, J. L., Gethyn T. J., Wilhelm, K., Zehnder, A.: Observing Photons in Space - A Guide to Experimental Space Astronomy. Springer Science+Business Media, New York (2013)
4. Lèna, P. et al.: Observational Astrophysics. Springer-Verlag Berlin Heidelberg (2012)
5. Schönfelder, V.: The Universe in Gamma Rays. Springer-Verlag Berlin Heidelberg, (2001)
6. Trumpër, J.E. and Hasinger, G.: The Universe in X-Rays, Springer-Verlag Berlin Heidelberg (2008)

---

[8] http://www.mpe.mpg.de/eROSITA.
[9] http://science.nasa.gov/missions/iue/.

Chapter 12
Gamma Ray Astronomy

The domain of gamma-rays is very extended and several observational techniques are needed to cover the range from the MeV region to the TeV region. The low energy side (up to tens GeV) is the domain of space based observatories. The very low energy gamma rays, up to a few MeV, can be detected with scintillator systems. The region up to a few tens MeV uses Compton telescopes, relying on the scattering of incoming photons. The region from tens MeV and above uses the conversion of gamma rays into pairs and the tracking of the produced particles, coupled to the energy measurement. The typical configuration uses a pair conversion system coupled to a tracker and an electromagnetic calorimeter. The high energy side, above tens GeV, is characterized by low fluxes and requires large collection areas. Instrumentation is ground based and relies on the production of electromagnetic showers in the atmosphere. Relativistic charged particles inside the shower produce Cherenkov light with an amount proportional to the energy of the primary particle, that is collected using large reflectors equipped with photon detectors. The Imaging Air Cherenkov Telescope technique, that allows the reconstruction of the shower morphology, is presented. The problem of discriminating the showers formed by gamma rays from the electromagnetic sub-showers in showers produced by cosmic rays is discussed. Another technique for high energy gamma detection uses water Cherenkov reservoirs equipped with photon detectors. The ground based and space based facilities (in operation or future) are presented.

12.1 Gamma Ray Sources

The region of gamma rays is conventionally divided into different energy regions [8, 10]:

- *High energy* (HE): from 1 MeV to about 30 GeV
- *Very High Energy* (VHE): from 30 GeV to 30 TeV
- *Ultra High Energy* (UHE): above 30 TeV

© Springer International Publishing Switzerland 2017
R. Poggiani, *High Energy Astrophysical Techniques*,
UNITEXT for Physics, DOI 10.1007/978-3-319-44729-2_12

The first region is covered using space based instruments. The second region is investigated using ground based Cherenkov arrays. The last region is the domain of ground based air shower arrays and fluorescence detectors.

Black body sources in thermal equilibrium are not able to produce gamma rays due to the temperatures that would be required, larger than billions K. The gamma rays are emitted in regions with relativistic particles that undergo a variety of processes [8, 10]: *synchrotron radiation* and *bremsstrahlung* from the acceleration of charged particles in a magnetic and electric fields; inverse Compton scattering; annihilation; nuclear transitions. In addition, gamma rays can be emitted in the collision of cosmic rays with matter encountered during the travel to the observer and in explosive phenomena, such as Gamma Ray Bursts (GRB) [2]. Due to the neutrality of photons, the observed radiation can be associated to the emitting source, unlike cosmic rays. A charged particle with mass m in a magnetic field B emits by the process of *synchrotron radiation*, whose spectrum is proportional to $\gamma^2 B^2 m^{-2}$, within a narrow emission angle, $\sim\gamma^{-1}$. The motion of a charged particle in an electric field, as the field of an atomic nucleus, produces radiation by *bremsstrahlung*. Gamma rays are the product of the collisions of relativistic particles. Low energy photons that scatter onto energetic particles gain energy, in the process of *inverse Compton scattering*, the reverse of the standard Compton process discussed in Chap. 2. The average energy gain is proportional to $(\beta\gamma)^2 E$, where E is the photon energy: low energy photons can be promoted to X-rays or gamma rays through the interaction with energetic electrons. Several *nuclear transitions* of nuclear levels are in the high energy photon region, more precisely in the MeV region. The excitation of some level and the subsequent disexcitation or decay is a possible source of gamma rays. Several elements have transitions in the MeV region: ^{26}Mg* ^{12}C* ^{16}O* ^{26}Al. Another source of gamma rays is given by the *decay of neutral pions* produced in the collision of cosmic rays with matter. The decay produces two photons, with an energy distribution centered at one half of the pion mass, about 70 MeV. The electron-positron *annihilation* in two photons, the inverse of the pair production process described in Chap. 2, produces two photons at 511 keV. Gamma rays are also expected as the product of self annihilation of a dark matter candidate, the neutralino (Chap. 16).

The fluxes of gamma ray photons are very low [8, 10]. The closest gamma ray sources have fluxes of the order of 10^{-2} to 10^{-3} photons cm^{-2} s^{-1} MeV^{-1}, that demand long observation times and large collecting areas. The effective area is often the area of the detector, since it is not possible to reflect or focus the gamma rays as in optical telescopes. The detectors should be thick enough to provide a high stopping power.

The flux densities of some gamma ray sources of interest are reported in Table 12.1.

In the following, we will present different techniques to reconstruct the energy, the direction and the arrival time of the gamma rays. The direct detection of low energy photons, with energies below some tens GeV, is performed in space. The detection of gamma rays with high energies is indirect and is performed by detecting the products of their interaction with the atmosphere. The recent instrumentation developments have lead to an overlap between the two regions.

Table 12.1 Flux density of X-ray sources [2]

Source	Flux density (photons/cm^2 s keV)	Energy (MeV)	Source type
Crab	6×10^{-11}	100	SNR+pulsar
Crab	2×10^{-13}	1000	SNR+pulsar
Geminga	3×10^{-11}	100	Pulsar
Geminga	6×10^{-13}	1000	Pulsar
Cyg X-1	1×10^{-5}	1	X-ray binary
3C 273	1×10^{-8}	100	Active galaxy
3C 273	3×10^{-11}	1000	Active galaxy

12.2 Low Energy Gamma Rays: Direct Detection in Space

We will start the discussion with the historical first instruments in space [4, 8, 9], consisting of a detector or a combination of detectors equipped with shielding, to discuss the problems related to the operation in space. The *direct detection of gamma rays* has to face different sources of background. The background in space based gamma ray telescopes is produced by the natural radioactivity of the instrument assembly and by the interaction with the cosmic rays [4, 7–9]. The reduction of the background requires a careful choice of the materials composing the gamma ray telescope. The interactions of cosmic rays with the instrument produces the excitation of nuclei and spallation products, whose decay produces gamma rays, either prompt or retarded, and neutrons. Neutrons produce radioactive materials by the process of activation. The background in the detector can be split into the *neutral particles background* and the *charged particle background*. The neutral particles are unwanted gamma rays not related to the astrophysical source of interest and neutrons. The decay of internal radioactivity and the activation of the materials in the satellite set the lower limit to the background of gamma rays, the *internal background*. The amount of passive material should be minimized. The other background is called *external background* and summarizes the byproduct of several physical processes. For balloon operated instruments, they include the interaction with the gamma rays and neutrons produced by incident cosmic rays in the residual atmosphere. The external background comprises also the astrophysical background, that is a topic of intrinsic interest. The charged particle background is due to cosmic rays crossing the detection system, leaving a clear signature, an ionization trail. This background is usually reduced enclosing the gamma ray detector inside a *shield* made of a plastic scintillator (with a typical thickness of the order of centimeters) that has an high detection efficiency for charged particles. The signal from the scintillator shield is used as a *veto* signal to perform an *anticoincidence*: events that show the signature of a signal in the shield are rejected. The plastic scintillator shield is an example of background suppression with *active shielding*. Plastic scintillators have a low stopping power for gamma rays and are not suitable to build collimators, that require scintillators with high atomic numbers, such as the inorganic crystals. A common solution is the *passive shielding* of a gamma ray detector by a collimator made of highly absorbing materials. However, the passive shield is a source of unwanted gamma rays produced by the interaction with the cosmic rays. The discrimination between gamma rays and neutrons relies on the different mechanism of the scintillation process, involving electrons for the electromagnetic interaction of gamma rays and protons for the nuclear interaction of neutrons. The choice of scintillating materials where the decay time of the pulse is different for different incident particles is used to identify them through the technique of *Pulse Shape Discrimination*, discussed in Chap. 6.

The different detection strategies for gamma ray detection in space have been pioneered by the *Compton Gamma Ray Observatory (CGRO)*, that has investigated gamma rays in different energy ranges from 1991 to 2000 [4, 5, 8–10]. The Compton Observatory consisted of four different instruments:

- The Oriented Scintillation Spectrometer Experiment (OSSE), sensitive in the region from 0.1 to 10 MeV
- The Compton Telescope (COMPTEL), for the energy interval between 1 and 30 MeV
- The Energetic Gamma Ray Experiment Telescope (EGRET), for the investigations of gamma rays with energies from 20 MeV to 30 GeV
- The Burst and Transient Source Experiment (BATSE), a full sky monitor for gamma ray transients, with the Large Area Detector (LAD) and the Spectroscopy Detector (SD), that covered together the interval from 15 keV to 1 MeV

The main features of the Compton Gamma Ray Observatory instruments are summarized in Table 12.2. The different technologies adopted in the instruments will be used as a guideline in the discussion.

12.2.1 Scintillator Systems

The OSSE instrument is an example of a scintillator systems operating in space [4, 5, 8, 9]. OSSE consisted of four separate detectors equipped with collimators. Each detector was a phosphor-sandwich or *phoswich* combination of two different

Table 12.2 Parameters of the Compton Gamma Ray Observatory instruments (based on [8])

Parameter	OSSE	COMPTEL	EGRET	BATSE LAD	BATSE SD
Energy range (MeV)	0.1–10	1–30	20–30000	0.03–1.9	0.015–110
Energy resolution (%)	4–12	6–9	20	20–32	6–8
Effective area (cm^2)	600–2000	30	1200–1600	500–1000	50–100
Position uncertainty (arcmin)	10	10	5–10	60	Not available
Field of view	$4^0 \times 11^0$	3 sr	0.6 sr	4π	4π

scintillators, a NaI(Tl) crystal on top of a CsI(Na) crystal and a tungsten collimator. The detector assembly was actively shielded with a layer of NaI(Tl) for anticoincidence. The phoswich assembly used the technique of pulse shape discrimination. The gamma rays coming from the collimator bottom were rejected relying on the higher probability of interacting with the CsI(Na). The scintillators and photomultiplier tubes were calibrated with a radioactive source of ^{60}Co and a reference LED on board.

The BATSE instrument was composed of eight modules at the corners of the CGRO satellite, that provided the measurement of the energy and the position of gamma ray transients [4, 5, 8, 9]. Each module consisted of a Large Area Detector (LAD) for the measure of the flux and direction of sources and a Spectroscopy Detector (SD) for the securing of the energy spectrum. The LAD was a NaI crystal enclosed in a lead/tin passive shield and equipped with a plastic scintillator shield for anticoincidence. The SD was a NaI(Tl) crystal with an energy resolution of the order of a few percent. BATSE has provided the first map of the Gamma Ray Burst distribution.

12.2.2 Compton Telescopes

The absorption coefficient of the typical materials for gamma ray detectors is smaller in the energy region from a few MeV to some tens MeV, where the dominant effect is Compton scattering, as discussed in Chap. 2. A *Compton telescope* (Fig. 12.1) consists of two plane detectors, often scintiilators, the scatter detector D_1 and the absorption detector D_2, separated by a distance of the order of a few meters [4, 5, 8, 9]. The scatter detector is made with a material with low atomic number, while the absorption detector is based on a material with a high atomic number to maximize the absorption probability. The Compton telescope has a large field of view, of the order of one sr, that makes it an ideal instrument for mapping the sky. The two detectors have the capability of reconstructing the interaction point.

Fig. 12.1 Layout of a Compton telescope

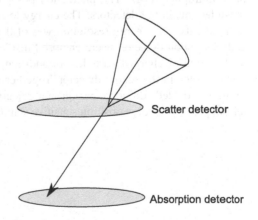

Scatter detector

Absorption detector

A gamma ray with energy E incident on the instrument deposits the energies E_1 and E_2 in the scatter and absorption detectors. The direction of the scattered gamma ray is reconstructed from the positions of the interaction in the two detectors: it lies on a circle described by a cone with an opening angle equal to the scattering angle around the direction of the scattered gamma ray. Assuming that the scattered gamma ray is completely absorbed in the second detector, the energy of the original gamma ray is:

$$E = E_1 + E_2 \tag{12.1}$$

while the scattering angle ϕ fulfills:

$$\cos\phi = 1 - \frac{m_e c^2}{E_2} + \frac{m_e c^2}{E_1 + E_2} \tag{12.2}$$

The determination of the direction of the incident gamma ray requires the measurement of the direction of the scattered photon. The scatter and absorption detectors can be segmented into smaller sections with separate readout to reconstruct the position of the interaction. An alternative approach is the use of the Anger camera technique, measuring the light produced in the scintillators with different photomultipliers and reconstructing the position of the interaction by their relative intensity. The background is reduced by the measurement of the Time of Flight of the scattered gamma ray from the scatter to the absorption detector. Only events where the photons cross the scatter detector before hitting the absorption detector are accepted. The background is strongly reduced if the position of the primary gamma ray is determined on the event circle: the track of the scattered electrons must be measured. The energy resolution of a Compton telescope depends on the energy resolution of the two detectors. The angular resolution is related to their position resolution and to the uncertainty on the scattering angle.

The COMPTEL instrument is an example of the Compton telescope technique. The scatter detector used some modules with liquid scintillator NE 213A, while the absorption one was made of NaI(Tl); the two detectors were separated by a distance of about 1.5 m. Both detectors were segmented into elements read by separated photomultiplier tubes. The interaction point was reconstructed using the relative pulse heights in the detectors. The energy losses were estimated by the sum of the light signals. The energy resolution was of the order of a few percent. The scatter and absorption detectors were enclosed inside a thin scintillator shell to reject the background of charged particles. In addition, the pulse shape discrimination was implemented in the scatter detector. Together with the measurement of the time of flight, it provided additional background rejection. The future evolution of Compton telescopes will involve position sensitive semiconductor detectors.

Fig. 12.2 Layout of
EGRET, a pair conversion
telescope

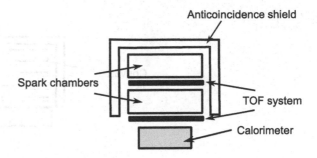

12.2.3 Pair Conversion Telescopes

The detection of photons with energies above 30 MeV is performed using *pair conversion telescopes*, based on instruments that provide the tracking of the electrons and positrons produced by the process of pair production [4, 5, 8–10] in a converter material. A *pair conversion telescope* consists of a pair tracking system combined with other instruments: a Time of Flight or a Cherenkov system to discriminate between upward and downward particles; a calorimeter for the measurement of the electromagnetic cascade; an anticoincidence shield for background rejection; a trigger telescope. The layout of a pair conversion telescope that will be discussed below, EGRET, is shown in Fig. 12.2.

The suitable converter materials have high atomic numbers. To reduce the scattering, thin layers of converter are interleaved with tracking detectors. The EGRET telescope is the prototype of a pair conversion telescope. EGRET used 28 thin tantalum sheets as converters and 36 spark chambers interleaved with the converters to measure the trajectory of electron and positrons and reconstruct the interaction point and the direction of the primary gamma ray. The charged particles traversed two scintillators that triggered the readout of the spark chambers and provided the time of flight. A calorimeter made of NaI(Tl) crystal was mounted below the spark chambers and provided the measurement of the impact position and the energy of the electron-positron pairs. The assembly was mounted inside a dome shaped anticoincidence shield made of a plastic scintillator to reject the charged particles. EGRET has built the first sky map in the gamma rays and a catalog of 271 sources. The energy resolution was about 20 %.

12.2.4 Space Based Observatories

The instrumentation of the CGRO observatory has opened the way to the present instruments in space. The tracker systems now rely on solid state technology.

Fig. 12.3 Layout of the
Fermi-LAT

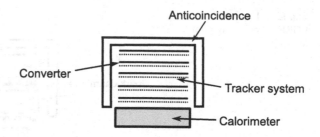

The *Astro-rivelatore Gamma a Immagini Leggero* (AGILE),[1] launched in 2007, is a telescope for the observation of gamma rays in the energy range from 30 MeV to 50 GeV, complemented by an X-ray imager for the region from 18 to 60 keV. The instrument consists of a silicon tracker, a CsI(Tl) calorimeter and an anticoincidence system and the X-ray system.

The *Fermi Large Area Telescope* (Fermi-LAT),[2] launched in 2008, is a telescope for the detection of gamma rays from 20 MeV to 300 GeV with a field of view of 2 sr (Fig. 12.3). The LAT instrument provides the measurement of the direction, the arrival time and the energy of each detected gamma ray. The LAT is a pair-conversion telescope, using the same strategy of EGRET, the combination of a conversion and tracking system with a calorimeter. The use of silicon microstrips for the tracking has canceled the necessity of a Time of Flight system. The effective collecting area of the Fermi-LAT is about 6500 cm^2 at 1 GeV, with an angular resolution of about 0.2° at high energy. The Fermi-LAT comprises 16 converter/tracker systems called towers. Each tower includes layers of silicon microstrip detectors interleaved with layers of tungsten converter sheets. The converter-tracker section is followed by the calorimeter, consisting of layers of CsI scintillators read with photodiodes. The tracker and the calorimeter have a thickness of ten radiation lengths. The LAT is equipped with a segmented anticoincidence shield made of plastic scintillator to reject the cosmic ray background using an on board trigger system. The Fermi-LAT has produced several catalogs of sources.

The Fermi satellite includes also the *Gamma-ray Burst Monitor* (GBM) for observations of transients in the energy range from 8 keV to 40 MeV.

The sensitivity of space based gamma ray instruments for continuum observations with an observing time of 10^6 s is reported in Fig. 12.4. The lower energy region needs a successor of COMPTEL. The high energy region has witnessed a steady improvement in sensitivity. The ground based MAGIC instrument, described in the following section, has a partial overlap with Fermi-LAT.

---

[1]http://agile.rm.iasf.cnr.it/.

[2]http://fermi.gsfc.nasa.gov.

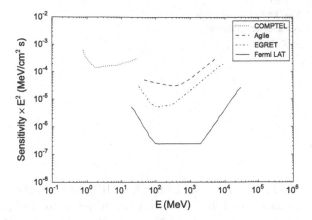

Fig. 12.4 Sensitivity of space based gamma ray instruments; data from [4]

12.3 High Energy Gamma Rays: Ground Based Detectors

Since the flux of gamma rays decreases with increasing energy, direct detection in space is not viable due to the large collection areas that would be required. The gamma rays with energies of tens GeV and above are detected using the atmosphere as an active medium, since they produce electromagnetic showers, as discussed in Chap. 2 [3–6, 8–10]. The shower particles are able to arrive at the ground level only for showers initiated by gamma rays with energies above 10 TeV (Fig. 3.4). The charged particles in the shower can be measured through their Cherenkov emission, even if they cannot be directly detected. As discussed in Chap. 3, the Cherenkov threshold for electrons/positrons is about 21 MeV at sea level and about 35 MeV at an altitude of 8 km. The emission angle of the Cherenkov radiation increases with the penetration in the atmosphere, varying from about 0.74° at 8 km to about 1.3° at the ground level. The Cherenkov radiation is focused into a ring with a radius of about 120 m at the ground level [3–5, 8–10]. The number of particles at maximum is proportional to the energy of the primary gamma ray. The process is the basis of the *Atmospheric Cherenkov Telescopes* (ACT).

The basic *Atmospheric Cherenkov Telescope* (ACT) consists of a large mirror with an optical photon detector at the focus (Fig. 12.5). The optical detector is usually a photomultiplier tube, the traditional transducer of high energy astrophysics. The mirror must have a large collecting area. The optical quality required for light collection is less stringent than that required in optical observations, making realization of large mirrors practical. The mirrors for ACT systems are built by assembling a set of spherical reflecting elements with the required focal length.

The ACT detectors have a limited duty cycle, smaller than 15 %, since the observations of Cherenkov light must be performed in dark sites during moonless nights. The ultimate background to observations is given by the night sky brightness. The integration time of the photomultiplier signal must be comparable with the duration of the Cherenkov flash, of the order of a few nanoseconds, to improve the signal to

Fig. 12.5 Layout of an
Atmospheric Cherenkov
Telescope

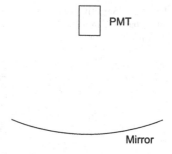

Fig. 12.5 Layout of an
Atmospheric Cherenkov
Telescope

noise ratio. As discussed by [3, 8, 10], the Cherenkov flash of a gamma ray with an energy of 1 TeV lasts for a few nanoseconds and produces a density of about 50 photons m^{-2} within 100 m about the shower axis. The contribution of the night sky is of the order of 10^{12} photons m^{-2} sr^{-1}. The choice of a small field of view, comparable with the angular size of the Cherenkov light, decreases the contribution of the night sky to a few photons m^{-2}.

12.3.1 Imaging Atmospheric Cherenkov Telescopes

The technique of *Imaging Atmospheric Cherenkov Telescopes* (IACT) allows the rejection of the signals produced by the unwanted electromagnetic component of hadron showers by using imaging. The photomultiplier at the focus of the mirror is replaced by an array of photomultipliers acting as a camera for the Cherenkov light. The Imaging Atmospheric Cherenkov Telescopes observe a shower triggered by a gamma ray with an array of reflectors. The single telescope is triggered by a signal in a few photomultipliers within a time window of a few tens of nanoseconds. The array of telescopes is triggered by the coincidence of the triggers in two or more single instruments. The coincidence of different instruments ensures a better rejection of the background and reduces the energy threshold.

The Cherenkov radiation travels as an almost flat front with a thickness of a few nanoseconds in time. The combination of the signals from different arrays is used to estimate the direction and the energy of the shower. The detection of the Cherenkov radiation of the charged particles inside a shower initiated by gamma rays has to face the background of electromagnetic subshowers produced inside the hadron showers triggered by cosmic rays. The ratio of the incidence rates of the gamma ray to cosmic rays is of the order of 10^{-4}–10^{-3} [10]. The discrimination between the gamma ray induced shower and the electromagnetic component of the hadron showers is performed using the different spatial morphology of the two cascades. The secondary products inside an hadron shower are generally emitted at larger angles than those of the components of an electromagnetic cascade (Fig. 12.6).

The distribution of the Cherenkov light projected onto the detector is larger for the electromagnetic subshowers in the hadron cascade than for the pure electromagnetic

Fig. 12.6 Schematic view of a two showers initiated by particle with identical energy, a gamma ray (*left*) and a cosmic ray (*right*)

showers. The image of the electromagnetic shower at the detector is an ellipse, whose major axis is the projection of the shower direction on the detector plane. The radial distribution is decreasing smoothly for the electromagnetic members of the hadron cascade, while it is quickly dropping at large radii for pure electromagnetic showers. An additional criterion is the larger spread of the Cherenkov light pulse of the electromagnetic component of the hadron cascade compared to the gamma ray triggered shower. The angular resolution of IACT systems is of the order of a fraction of degree, while the energy resolution ranges from 30–40 % with a single detector to 10–15 % with more detectors [8, 10]. Electrons incident on top of the atmosphere produce an electromagnetic shower that is indistinguishable from a gamma ray shower.

Different facilities have explored the techniques of Cherenkov detection. The *Whipple* telescope used a single reflector with a diameter of 10 m. The showers induced by gamma rays were discriminated against the showers induced by cosmic rays using the difference in the morphology. Whipple has provided the first evidence of TeV radiation from the Crab.

The *High Energy Gamma Ray Astronomy* (HEGRA) experiment[3] at La Palma consisted of six telescopes with photomultiplier arrays at the focus and provided a stereoscopic imaging of the shower.

The *Very Energetic Radiation Imaging Telescope Array System* (VERITAS)[4] in USA is an array of four reflectors with an aperture of 12 m with a field of view of a few degrees. The instrument measures the gamma rays with an energy larger than

---

[3]https://www.mpi-hd.mpg.de/hfm/HEGRA/HEGRA.html.
[4]http://veritas.sao.arizona.edu/.

50 GeV with an angular resolution of the order of 0.1^0 and an energy resolution of 10–15 %.

The *High Energy Stereoscopic System* (HESS)[5] in Namibia is made of four telescopes with an aperture of 13 m at the edges of a square with a side of 120 m and a central telescope with an area of 600 m². The telescopes have a fast slew rate for pointing. The reflectors are equipped with camera with fine pixelation that allows good discrimination of electromagnetic and hadron showers. The instrument has an angular resolution of a fraction of degree.

The *Major Atmospheric Gamma-ray Imaging Cherenkov* (MAGIC)[6] at La Palma is an array of two telescopes with an aperture of 17 m and a field of view of a few degrees; the reflectors can be pointed in the direction of possible transients in less than one minute. The instrument has a low energy threshold at about 25 GeV.

The future evolution of ground based arrays is the *Cherenkov Telescope Array* (CTA).[7] The instrument will extend the traditional region of operation of Cherenkov instruments, lowering the threshold at about 20 GeV. The array will use a combination of reflectors with different sizes, with apertures from a few meters to tens meters, for the different energy regions. The low energy instrumentation (20–200 GeV) will use four reflectors with an aperture of 23 m. The instrumentation for the medium energy range (100 GeV–10 TeV) will consists of 40 telescopes with an aperture of 12 m. The high energy instrumentation (10–300 TeV) will use seventy reflectors with an aperture of 4 m.

12.3.2 Air Shower Arrays

The gamma rays with energies above 10 TeV produce secondary particles that reach the ground level where they are detected with *Air Shower Arrays* [3, 4, 8–10], arrays of detectors distributed over large areas, analogous to the technique used for cosmic ray detection (Chap. 13). The direction of the primary gamma ray is estimated by the arrival times in the different detectors, while the energy is estimated by the total number of detected particles. The air shower arrays can operate during the daytime and do not have the duty cycle limitations of the Cherenkov telescopes. An additional advantage is the large field of view, compared to the few degrees of the Cherenkov systems.

The *Milagro* detector[8] at Los Alamos was a water Cherenkov detector sampling the showers, consisting of a center tank with an area of about 5000 m² and additional water tanks distributed over an area of about 40000 m². The detection of the Cherenkov light used the large Cherenkov angle in water by placing photomultiplier tubes at a depth that is about one half of the spacing to detect all particles in the

---

[5]https://www.mpi-hd.mpg.de/hfm/HESS/.

[6]https://magic.mpp.mpg.de/.

[7]https://portal.cta-observatory.org/Pages/Home.aspx.

[8]http://umdgrb.umd.edu/cosmic/milagro.html.

Fig. 12.7 Sensitivity of ground based gamma ray observatories; data from [1]

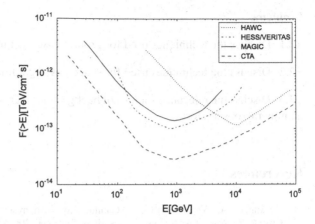

shower. The photomultiplier tube system was able to detect also the gamma component, since it was positioned after four radiation lengths of water. An additional layer of photomultiplier tubes was placed at about 16 radiation lengths to measure the hadronic component.

The *High-Altitude Water Cherenkov Observatory* (HAWC)[9] in Mexico consists of 300 Cherenkov tanks with a diameter of 7.3 m. The readout of each tank is performed with four upward facing photomultiplier positioned at the bottom. The events are reconstructed by the fit of the spatial charge distribution to the primary particle trajectory projected onto the array. Hadronic showers are separated from the electromagnetic showers since they deposit energy with a patch like structure, while gamma rays show a clear core structure. An additional signature of the hadronic events is the presence of muons, that are identified by the large amount of Cherenkov light.

The *Astrophysical Radiation with Ground-based Observatory* (ARGO)[10] in Tibet used a single layer of resistive plate chambers over an area of about 10^4 m^2.

The sensitivity of ground based gamma ray observatories is summarized in Fig. 12.7, after [1]. The sensitivities are measured as the minimum intensity that can be detected at 5 σ, with one year of operation for the air shower arrays and 50 h for the Cherenkov instruments.

---

[9]http://www.hawc-observatory.org/.
[10]http://argo.na.infn.it/.

Problems

12.1 Discuss the techniques used for ground based gamma ray observatories.

12.2 Discuss the techniques used for space based gamma ray observatories.

12.3 Discuss the relevance sources of background in ground based and space based gamma ray observatories.

References

1. De Angelis, A.: (Very)-High-Energy Gamma-Ray Astrophysics: the Future. arXiv:1601.02920
2. Cox, A. N.: Allen's Astrophysical Quantities. Springer (2002)
3. Grieder, P. K. F.: Extensive Air Showers. Springer (2010)
4. Grupen, C. and Buvat, I: Handbook of Particle Detection and Imaging. Springer-Verlag Berlin Heidelberg (2012)
5. Huber, M. C. E., Pauluhn, A., Culhane, J. L., Gethyn T. J., Wilhelm, K., Zehnder, A.: Observing Photons in Space - A Guide to Experimental Space Astronomy. Springer Science+Business Media, New York (2013)
6. Lèna, P. et al.: Observational Astrophysics. Springer-Verlag Berlin Heidelberg (2012)
7. Olive, K.A. et al. (Particle Data Group): Chin. Phys. C **38**, 090001 (2014)
8. Schoönfelder, V.: The Universe in Gamma Rays. Springer-Verlag Berlin Heidelberg (2001)
9. Spurio, M.: Particles and Astrophysics - A Multi-Messenger Approach. Springer International Publishing, Switzerland (2015)
10. Weekes, T.: Very High Energy Gamma-Ray Astronomy. Institute of Physics Publishing, Bristol and Philadelphia (2003)

Chapter 13
Cosmic Ray Astronomy

Observatories for cosmic rays are ground based or space based. As discussed for gamma rays, the low energy side is the domain of balloon borne or space based observatories, that show close similarities with the instrumentation at colliders. The cosmic ray detection and energy measurement is performed with a combination of magnetic spectrometers for momentum measurement, calorimeters for energy measurement and systems for particle identification. Due to the decreasing rates with increasing energy, cosmic rays at higher energies are measured with ground based arrays consisting of a large number of detectors distributed over extended areas, the Extensive Air Shower arrays, that include scintillators, water Cherenkov detectors and so on. At the highest energies, the detection of fluorescence of nitrogen molecules is used. The air shower and fluorescence detection techniques are combined in hybrid detectors.

13.1 Cosmic Rays Sources Fluxes

The cosmic ray showers have been discovered by Auger in 1938 and for two decades at least have been the natural accelerators of the new born high energy physics. Several elementary particles have been discovered in cosmic rays, among them the positron, the muon, the charged pions, the kaons, just to name a few. After the dawn of particle physics at accelerators, the cosmic rays studies have continued to provide information about the physical processes in the realm of ultra high energies.

The energy spectrum of the primary cosmic rays shown in Fig. 13.1 is the combined result of a large number of experiments. The range of possible energies span several orders of magnitude in energy, that require different observational techniques.

The flux of primary cosmic rays is steeply decreasing with energy. The integrated flux ranges from thousands particles/s/m^2 at hundreds GeV, to some particle/year/m^2 at 10^{15} eV to a mere 1 particle/century/km^2 at 10^{20} eV. The direct detection of cosmic rays, performed in space, is possible only in the low energy region, while the highenergy primaries are measured through the cascade of particles they produce in

© Springer International Publishing Switzerland 2017
R. Poggiani, *High Energy Astrophysical Techniques*,
UNITEXT for Physics, DOI 10.1007/978-3-319-44729-2_13

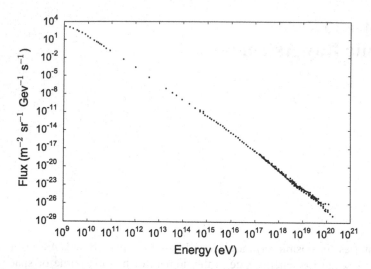

Fig. 13.1 Cosmic ray spectrum; data from http://www.physics.utah.edu/~whanlon/spectrum.html

the atmosphere. The spectrum shows different regimes, marked by different slopes. The first transition point, called *knee*, occurs at an energy of about 3×10^{15} eV. The second transition, called *ankle*, occurs at a few times 10^{19} eV. Several particles have been detected above the GZK cut-off (Chap. 3). Cosmic rays can explore the energy region well above the LHC one. The cosmic ray physics is coming back to play a strong role in the high energy investigations. Cosmic rays at the lower energies are believed to be associated to supernovae, while those at energies above about 10^{18} eV are probably produced in extragalactic sources.

The study of the chemical composition of the cosmic rays is essential to understand the astrophysical sources producing them and the interactions during their propagation. The observations of composition have been performed outside the atmosphere or at the altitude of balloons at least with *direct detection* methods, for energies up to 10^{14} eV. At larger energies, the low flux dictates the use of *indirect detection* methods, with large area instruments that are ground based and rely on the shower of particles produced by the interaction of the incident cosmic ray with the atmosphere.

13.2 Low Energy Cosmic Rays: Direct Detection in Space

The budget of cosmic rays comprises protons and nuclei, but also electrons and antiparticles. The *direct measurements of the cosmic rays* with energies smaller than about 100 TeV are secured with balloon borne or space based observatories [1–3, 7]. The facilities provide the measurement of the flux and of the composition of the cosmic rays. The energy region investigated by the direct measurements deals

Fig. 13.2 Layout of the IMP
experiment

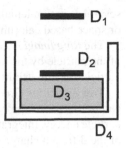

with the cosmic rays that, according to current theoretical models, are related to
supernovae.

The *IMP* experiment is an example of the first generation of cosmic rays observatories in space [4, 7]. The instrument consisted of four detectors D_1, D_2, D_3, D_4
(Fig. 13.2). The D_1, D_2 were thin Lithium drifted semiconductor detectors that
defined the acceptance of the telescope, D_3 was a CsI(Tl) scintillator and D_4 a plastic
scintillator acting as an anticoincidence shield. The interesting events corresponded
to the crossing of the first three detectors D_1, D_2, D_3 detectors, but not of D_4: the
signature of the event was $D_1 D_2 D_3 \overline{D_4}$. The behavior of the signal in the $D_1 + D_2$
pair against the total signal $D_1 + D_2 + D_3$ is a probe of the ionization energy loss
against the energy and discriminates between particles with different atomic number.

The present instrumentation strategies of space based observatories have been
pioneered by *PAMELA*, launched in 2006, and *AMS-02*, launched in 2011, that aim
to measure the energy spectrum and the composition of the cosmic rays, but also to
search for antimatter particles [2, 3, 7]. Both observatories have included *magnetic
spectrometers* in the instrument. There is no detection of neither antiproton nor
antihelium nuclei, but the detection of a positron excess has thrown a bridge towards
the astrophysics of dark matter (Chap. 16). The detection and identification of cosmic
rays requires the measurement of the charge Z, the mass A and the energy E. The
measurement of the energy is performed using a combination of techniques: *magnetic
spectrometers* for the low energy region, up to TeV energies, and *calorimeters* above.
The triggering is provided by the Time of Flight (TOF) systems. The instrumentation
is redundant in the choice of the detectors to measure the energy of the cosmic ray
over the largest possible range and to discriminate between positrons and protons,
and nuclei with different charges.

The magnetic spectrometers are the combination of the magnetic field produced
by a solenoid with a tracking system. The requirement of the precise reconstruction
of the tracks sets the upper limit to the measurable rigidity, that is smaller than a few
TV. The trajectory of the charged particle in the magnetic field allows to reconstruct
the charge and its sign, to identify antimatter particles.

The measurement of the energy is performed with calorimeters, similar to the
calorimeters of high energy physics. The shower produced in the material must be
contained, at least a good fraction of it. The prescription of having a thickness of

several radiation lengths or interaction lengths cannot always be fulfilled in balloon or space based calorimeter, due to the mass issues.

The *Ring Imaging Cherenkov Detectors* (RICH) are used the measure the velocity of the particle by the radiation pattern of the Cherenkov photons. The charge of the particle is estimated by the intensity of the Cherenkov radiation.

The *Transition Radiation Detectors* (TRD) are used to measure the relativistic gamma factor of the particle, thus they provide a non calorimetric estimate of the particle energy. Integrated with other detectors, the TRDs contribute to the separation of the different charges.

13.2.1   Modern Observatories

The balloon experiments have a long tradition in cosmic ray physics and are still used [2, 3]. The present balloons are able to carry payloads of a few tons at an altitude of some tens km. The *Cosmic Ray Energetics and Mass Balloon Experiment* (CREAM)[1] has investigated the charge range up to 28. The experiment measured the energy of particles up to 10^{14} GeV using two instruments, a W/scintillating fiber calorimeter with a thickness of 20 radiation lengths and a TRD system. The redundant strategy allowed the intercalibration. The *Advanced Thin Ionization Calorimeter* (ATIC) instrument[2] has used a BGO calorimeter with a thickness of 20 radiation lengths and a silicon array to resolve the trajectory and an hodoscope of scintillators between the two. The *Transition Radiation Array for Cosmic Energetic Radiation* (TRACER)[3] has used a TRD to measure the gamma factor of the particles, coupled to a pair of scintillators to measure their charge. The *Japanese American Collaborative Emulsion Experiment* (JACEE)[4] has used emulsions as calorimetric media.

The *Payload for Antimatter Matter Exploration and Light-nuclei Astrophysics* (PAMELA)[5] is satellite-born experiment to measure the composition of the cosmic rays up to the TeV region. The main elements of the instrument are shown in Fig. 13.3. The TOF system consists of three layers of scintillators S_1, S_2, S_3 and it is used as a trigger and to measure the ionization energy loss of particles, allowing also the rejection of particles that enter the instrument from the bottom side. The magnetic spectrometer has a field of about 0.4 T and is equipped with a silicon tracking system with a position resolution of a few μm. The spectrometer is followed by a sampling Si/W electromagnetic calorimeter with a thickness of about 16 radiation lengths and an energy resolution of some percent. The calorimeter provides the measurement of the energy and the lepton/hadron separation. It is followed by a scintillator and a neutron detector, to improve the lepton/hadron separation.

---

[1] http://cosmicray.umd.edu/cream/.
[2] http://atic.phys.lsu.edu/.
[3] http://tracer.uchicago.edu/.
[4] http://marge.phys.washington.edu/jacee/.
[5] http://pamela.roma2.infn.it/index.php.

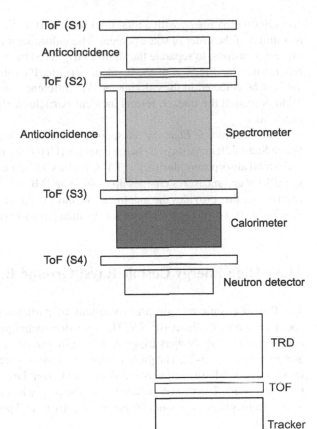

Fig. 13.3 Layout of the PAMELA experiment

Fig. 13.4 Layout of the AMS-02 experiment

The *AntiMatter Spectrometer* (AMS-02)[6] (Fig. 13.4) is installed on the International Space Station. It uses a permanent magnet enclosing the Central Tracker, based on silicon microstrip sensors, that measures the curvature of the track of charged particles, with a maximum measurable rigidity is 2 TV. The alignment of the tracker modules is controlled by an on board system. The TOF system consists of two double layers of scintillators above and below the magnet and provides also the trigger, helping in discriminating between up and down going particles. The TRD is placed on the top side of the magnet. The electromagnetic calorimeter is a made of layers of lead

---

[6]http://www.ams02.org/.

and scintillating fibers, with a total thickness of 16 radiation lengths and an energy resolution of the order of some percent. The calorimeter reconstructs the showers in three dimensions, to separate the showers triggered by positrons and protons and to reconstruct the direction of the incident particle. Photons can produce pairs in the TRD or be detected in the calorimeter. An anticoincidence counter, a shell of scintillators around the tracker, rejects incident particles arriving at large angles acting as a veto.

The *Calorimetric Electron Telescope* (CALET)[7] is installed on the International Space Station. It consists of a charge detector (CHD), an imaging calorimeter (IMC) and a total absorption calorimeter (TASC). The CHD consists of two layers of plastic scintillator and measures charges up to 40. The IMC is a tungsten/scintillating fiber calorimeter that provides the identification and the direction of the incident particle. The TASC is a PbWO calorimeter for the measurement of the energy.

13.3   High Energy Cosmic Rays: Ground Based Detection

The flux of cosmic rays drop to some tens of particles per square meter per year above an energy of about 10^{15} eV. The detection techniques is indirect, targeting the components of the showers triggered by the interaction of the cosmic rays with the atmosphere [1, 2, 5–7]. The ground based instruments are the *Extensive Air Shower* (EAS) arrays whose elements are distributed over large areas, up to hundreds or thousands km^2. The shower initiated by a cosmic ray includes an hadron component, an electromagnetic component (photons, electrons and positrons), muons, neutrinos (Chap. 3).

13.3.1   Extensive Air Shower Arrays

The Extensive Air Shower arrays investigate the showers started by cosmic rays with energies in the interval from 10^{14} to 10^{17} eV [2, 3]. The EAS experiments have shown that the arrival direction of the cosmic rays in this energy interval is isotropic, making the identification of the original source unfeasible. They EAS arrays are made of a combination of detectors distributed over large areas (Fig. 13.5). The detectors used in the array combine different technologies to measure the electromagnetic and muon components.

The performance of the arrays is characterized by the quantity $\Phi_t A \Delta \Omega T$, where Φ_t is the flux of particles above a chosen energy E, T the observation time, $\Delta \Omega$ the field of view of the array and A the area. The spacing of the single detectors is of the order of some tens meters and increases with the energy of the cosmic ray. The direction and the energy of the incident cosmic ray is reconstructed assuming

---

[7]http://calet.phys.lsu.edu.

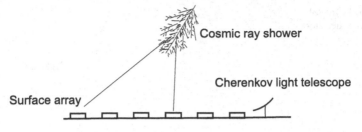

Fig. 13.5 Layout of an EAS array

axial symmetry of the products. The relativistic charged particles in the array emit by Cherenkov effect. The ideal EAS array should measure both the electron and muon component to reconstruct the energy of the primary cosmic ray.

The electron and muon components electrons propagate as wavefronts orthogonal to the shower axis. Both are thin layer of particles, with a thickness that increases moving far from the shower axis. The muon wavefront precedes the electromagnetic one. The structure of the shower allows to reconstruct the arrival direction of the cosmic ray by the measurement of the arrival times of the front. The uncertainty on the direction is related to the time resolution of the array detectors, being of the order of a fraction of degree. The lateral distribution of the shower is estimated by a fit with a modified version of the NKG function (Eq. 3.15), keeping the age s of the shower as a free parameter or using a guess value. The energy of the shower or total number of particles is estimated from the integral of the radial distribution.

The energy region close to the knee feature in the spectrum has been investigated by several past instruments that have been the models for the present detectors. The array at Volcano Ranch consisted of twenty plastic scintillators with an area of 3 m^2 each that sampled a surface of 8 km^2. The *Extensive Air Shower on Top* (EAS-TOP)[8] on top of Gran Sasso used 35 scintillator modules and a central muon-hadron calorimeter made of iron and streamer tubes. The *KArlsruhe Shower Core and Array Detector* (KASCADE) experiment[9] includes an array of 252 detectors spread over an area of $4 \times 10^4 \text{ m}^2$; each detector consists of a liquid scintillator for electron/γ detection on top of lead and iron slabs. The outer units of the array have also scintillators for muon detection positioned below the metal slabs. The KASCADE-Grande is an upgrade of the original configuration that covers an area of about $5 \times 10^5 \text{ km}^2$. Two other experiments combine the surface detectors with underground detectors for muons. The *Yakutsk* array[10] covers an area of 18 km^2. The charged particles are detected by scintillators on the surface, while muons are detected by scintillators underground. The experiment uses also photomultiplier tubes to detect the Cherenkov light emitted by charged particles. The *Chicago Air Shower Array* (CASA) with the

[8]http://web.lngs.infn.it/lngs_infn/contents/lngs_en/research/experiments_scientific_info/experiments/past/eastop/.

[9]http://www-ik.fzk.de/KASCADE_home.html/.

[10]http://www.eas.ysn.ru/index.php.

Fig. 13.6 Layout of a
fluorescence detector

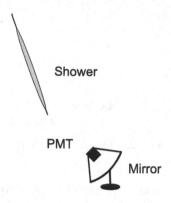

Muon Array (MA) covered an area of 2.23 km^2 and used a combination of array and
surface scintillators.

A byproduct of the development of the hadronic shower is the Cherenkov radiation
emitted by the relativistic particles, that can be detected with photomultiplier tubes.
The observation of Cherenkov radiation has been incorporated into the CASA-MIA
experiment with the *Broad LAteral Non-imaging Cherenkov Array* (BLANCA) and
the *Dual Imaging Cherenkov Experiment* (DICE).

13.3.2 Fluorescence Detection

The cosmic rays with energies above 10^{18} eV are called *Ultra High Energy Cosmic
Rays* (UHECR) and are probably produced by extragalactic sources. The EAS array
technique can be used also in this energy region [1, 2, 7]. An example of EAS array for
the measurement of UHECRs is *Akeno Giant Air Shower Array* (AGASA).[11] It used
111 scintillators distributed over 100 km^2 to sample the electromagnetic component
of the showers. The AGASA data do not show the evidence of the GZK cut-off. An
alternative approach, the *fluorescence detection*, relies on the excitation of nitrogen
molecules by the particles in the shower and the detection of the emitted light [1,
2, 7]. The lifetime of the excited states is of the order of ten nanoseconds. The
fluorescence emission is isotropic and has a maximum in the wavelength region
between 300 and 440 nm, with a light yield of 7 photons/MeV at 337 nm, in dry
air at 800 hPa, 293 K [6]. The amount of fluorescence light is proportional to the
number of charged particles in the shower. The isotropicity of the emitted light
allows the observation of distant showers. The detecting system consists of an array
of photomultiplier tubes packed as a camera and a reflecting mirror (Fig. 13.6). The
Fly's Eye experiment has pioneered the fluorescence detection technique using a
three dimensional configuration of detectors shaped as the eye of the fly.

---

[11] http://www-akeno.icrr.u-tokyo.ac.jp/AGASA/.

A single fluorescence instrument defines the shower-detector plane, two separate instruments define the axis of the shower as a *stereo reconstruction*. The angular resolution achieved with one and two instruments ranges from a few degrees to a fraction of degree. The total energy of the shower is proportional to the sum of the collected radiation. A drawback of the fluorescence detection technique is the low duty cycle: the faintness of the fluorescence signals allows observations only during moonless nights. The *HiRes Observatory*[12] was a system with two fluorescence detectors spaced by 12.6 km, each one consisting of several telescopes with individual mirrors and arrays of photomultiplier tubes.

13.3.3 Hybrid Detectors

The *hybrid detectors* combine the extensive air surface arrays with the fluorescence detectors, improving the detection capability and allowing a cross calibration of the instrumentation. The *Pierre Auger Observatory* (PAO)[13] in Argentina covers an area of 3000 km^2. It uses 1660 water Cherenkov detectors as surface detectors. Each tank has an area of 10 m^2 and is equipped with photomultipliers to measure the light produced by the charged particles. The shower core and the radial distribution are reconstructed using the same techniques used with cosmic rays at lower energies. The energy of the primary cosmic ray is estimated from the signal at a fixed distance from the shower axis. Four fluorescence detectors are placed on the borders of the surface array area. Each fluorescence detector includes six telescopes with 3.6 m × 3.6 m mirrors and arrays of photomultipliers. The two detection strategies allow to reconstruct the physical properties of the shower by comparing the signals. The shower direction is estimated with the fluorescence detector and the arrival time at the surface detectors. The energy of the primary cosmic rays is measured using the fluorescence detector system. For showers that are observed with the surface detectors and with the fluorescence detectors, the energy reconstructed with the fluorescence is compared with the estimation of the surface detectors at a fixed distance from the shower axis, to get a correction factor for all showers that are measured with the surface array. The AUGER experiment has observed a correlation of the highest energy cosmic rays with close extragalactic objects.

The *Telescope Array Observatory* (TA)[14] in USA covers an area of 762 km^2. It consists of three stations of fluorescence detectors with a spacing of about 30 km and about 500 scintillators as surface detectors covering an area of about 700 km^2.

---

[12]http://www.cosmic-ray.org/.

[13]https://www.auger.org/.

[14]http://www.telescopearray.org/.

13.3.4 Radio Detection of Showers

The hadronic showers initiated by cosmic rays produce radio emission that can be detected at some distance from the shower itself [1] and can overcome the limitations of the low duty cycle of the instruments relying on the detection of optical radiation, the Cherenkov telescopes and the fluorescence systems. The radio detection is performed in the region of tens MHz. The main processes leading to the generation of the radio bursts is the geo-synchrotron radiation of the electrons and positrons separated by the geomagnetic field. The amplitude of the radio signal at a characteristic core distance is proportional to the energy deposited in the atmosphere by the electromagnetic cascade. The technique is under development.

Problems

13.1 Discuss the techniques used for ground based cosmic ray observatories.

13.2 Discuss the techniques used for space based cosmic ray observatories.

13.3 Discuss the relevance sources of background in ground based and space based cosmic ray observatories.

References

1. Grieder, P. K. F.: Extensive Air Showers. Springer (2010)
2. Grupen, C. and Buvat, I: Handbook of Particle Detection and Imaging. Springer-Verlag Berlin Heidelberg (2012)
3. Huber, M. C. E., Pauluhn, A., Culhane, J. L., Gethyn T. J., Wilhelm, K., Zehnder, A.: Observing Photons in Space - A Guide to Experimental Space Astronomy. Springer Science+Business Media, New York (2013)
4. Longair, M. S.: High Energy Astrophysics: Volume 1. Photons, Particles and their Detection. Cambridge University Press (1992)
5. Lèna, P. et al.: Observational Astrophysics. Springer-Verlag Berlin Heidelberg (2012)
6. Olive, K.A. et al. (Particle Data Group): Chin. Phys. C **38**, 090001 (2014)
7. Spurio, M.: Particles and Astrophysics - A Multi-Messenger Approach. Springer International Publishing, Switzerland (2015)

Chapter 14
Neutrino Astronomy

Astrophysical neutrinos cover a wide range of energies. Due to the low interaction cross sections, neutrino detectors are built with large volumes of suitable materials. The chapter presents the techniques used for the detection of solar neutrinos, neutrinos from supernovae and high energy neutrinos. Neutrino telescopes for high energy neutrinos use optical modules installed in transparent media, such as water or ice, to record the Cherenkov light of the charged product of neutrino interactions.

14.1 Neutrino Fluxes

The neutrinos are related to high energy photons and to cosmic rays in a multimessenger connection. The low interaction cross sections of neutrinos allows them to probe physical processes and environments that are not accessible to electromagnetic probes, but makes their detection extremely difficult. The neutrino spectrum extends for more than ten orders of magnitude in energy (Fig. 14.1).

The low energy neutrinos, with energies of the order of MeV, include solar neutrinos and supernova neutrinos. The atmospheric neutrinos produced in the interaction of the cosmic rays with the Earth atmosphere have intermediate energies. The large detectors for the search of low energy neutrinos from the Sun and supernovae neutrinos are often used for the investigation of atmospheric neutrinos and the search for oscillations. The high energy neutrinos, above tens GeV, from extragalactic sources as the active galactic nuclei (AGN), are investigated with neutrino telescopes.

14.2 Neutrino Telescopes

The *neutrino telescopes* [1, 3, 5] investigate neutrinos with energies above some GeV. The small interaction cross sections demand the use of large volume detectors, presently at the kilometer scale. The detector can be underwater, underground or in

© Springer International Publishing Switzerland 2017
R. Poggiani, *High Energy Astrophysical Techniques*,
UNITEXT for Physics, DOI 10.1007/978-3-319-44729-2_14

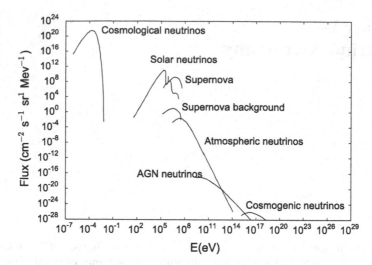

Fig. 14.1 Flux of astrophysical neutrinos; data from [2]

ice. The neutrinos are detected by detecting the secondary charged particles produced in their interactions with the material. The chosen media are transparent, often water or ice, and are equipped with arrays of light detectors to detect the Cherenkov radiation of the charged products. The number of detected photons and the arrival times are used to identify the type of neutrino, its energy and its direction. Large volumes of water or ice provide at the same time an active medium and a shield against the cosmic ray showers.

The neutrinos interact through neutral current (NC) interactions [2]:

$$\nu_l + N \rightarrow \nu_l + X \tag{14.1}$$

or charged current (CC) interactions:

$$\nu_l + N \rightarrow l + X \tag{14.2}$$

where l is a lepton, N a nucleon. The charged current cross section is proportional to the neutrino energy up to 10^{13} eV and has a value of 10^{-35} cm^2 at 1 TeV [4]. The *effective area* of a neutrino telescope is the product of its geometrical area and of several factors: the probability that a neutrino produces a muon, the efficiency of the muon detection, the neutrino absorption by the Earth. The typical values are in the range 10^{-1} to 10^3 m^2 above 1 TeV [4].

The neutrino telescopes in water or ice are arrays of *optical modules*, photomultiplier tubes mounted inside pressure tight and transparent spherical vessel together with the power supply, the electronics and LEDs for calibration. The optical modules record the arrival time and the amplitude of the Cherenkov pulse of the charged particles in the showers. The units are mounted along strings that are deployed in

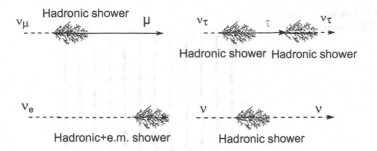

Fig. 14.2 Signatures of neutrino interactions

the medium at large depth, to block the contribution of the day light. The spacing between modules along a string is of the order of a few tens meters, while the strings are positioned at distances of about of 100 m. Water and ice are not equivalent. The absorption lengths of radiation is of the order of 100 m for ice and of 70 m for water. Sea water has a radiation background by bio-luminescence and a radioactive background from ^{40}K, while ice has a negligible radioactivity. The energy and the direction of the neutrino are reconstructed from the structure of the signals in the modules.

The signatures of neutrino events in a neutrino telescope are the presence of *showers* without muons or *tracks* by muons. The basic event shapes are summarized in Fig. 14.2. A charged current interaction can produce three different event morphologies. A muon neutrino produces a muon and an hadronic shower; a tau neutrino produces a τ that decays into a tau neutrino, leading to a pair of hadronic showers, the *double bang events*; an electron neutrino produces an hadronic shower and an electromagnetic shower. The neutral current interaction produces an hadronic shower only. A qualitative discrimination between the astrophysical neutrinos and the background is performed using an energy threshold, since the former have an harder spectrum than the latter. The muon neutrino has a muon track that allows to investigate also interaction events that occurred outside the sensitive volume. The astrophysical signature is an excess of events within a narrow angular region.

The lepton in the final state is directed as the initial neutrino within an angle of the order of $1.5^0/\sqrt{E(GeV)}$ [4]. Among all reactions, the clearest signature is the charged current interaction of the muon neutrino with a muon in the final state. The energy and the direction of the muon are reconstructed from the measurement of the emitted Cherenkov radiation. Differently from other telescopes, the *neutrino telescopes* observe the upward going muons produced in the charged current interaction of muon neutrinos. The upward direction ensures that the muon is coming from a neutrino, the only particle able to cross the Earth. The detectors are positioned at large depths to reduce the background of downward moving muons that could be interpreted as upward going muons. There is an additional source of background that cannot be avoided. Cosmic rays produce muons and neutrinos in the interaction with

Fig. 14.3 Schematic view of
the IceCube instrument

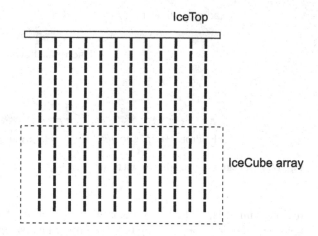

the Earth atmosphere (Chap. 3), the *atmospheric neutrinos*, that are mostly muon neutrinos from the decay of charged pions and kaons.

The neutral current interaction and the charged current interactions of electron and tau neutrinos do not have a muon in the final state, but only showers, electromagnetic or hadronic. The total amount of Cherenkov light is proportional to the energy of the shower, for events contained inside the detector the energy of the shower can be reconstructed with a precision of 10–20 % [4].

The first neutrino telescopes in water has been *BAIKAL*[1] at the Baikal Lake, with 200 optical modules at a depth of 1100 m. *Antarctic Muon And Neutrino Detector Array* (AMANDA)[2] in Antarctica has been the first neutrino telescope in ice with 677 optical detectors distributed over 19 strings. The *Astronomy Neutrino Telescope ans Abyss Environmental Research* (ANTARES) detector[3] is in the Mediterranean Sea, close to the French coast. The array of 12 strings comprises hundreds photomuliplier tubes installed at a depth of about 2500 m. ANTARES has access to the galactic center. The *IceCube* detector[4] at the South Pole is distributed over a volume of the order of km^3. More than 5000 photomultipliers modules have been installed in strings in ice at depths ranging from 1500 to 2500 m (Fig. 14.3). The observatory is completed by a cosmic ray detector on the surface, called IceTop. ANTARES and IceCube have a partial overlap in the coverage of the sky.

The future evolution of neutrino telescopes is *KM3NeT*,[5] a larger facility with a volume of several km^3, to be installed in the Mediterranean Sea; the foreseen detectors will be three dimensional, using arrays of photomultipliers.

---

[1] https://astro.desy.de/neutrino_astronomy/previous_projects/baikal/index_eng.html.

[2] https://astro.desy.de/neutrino_astronomy/previous_projects/amanda/index_eng.html.

[3] http://antares.in2p3.fr.

[4] http://icecube.wisc.edu.

[5] http://www.km3net.org/.

14.3 Solar Neutrino Detection

The solar neutrinos have low energies, of the order of MeV [1, 3, 5], and interact through the reaction:

$$\nu_e + {}^A_Z X \rightarrow {}^A_{Z+1} Y + e^- \tag{14.3}$$

that produce a transformation of a nuclide into a different one. The occurring of the reaction is marked by the appearance of new atoms of species Y, that must be separated by the atoms of species X with chemical extraction techniques. The species Y is easily targeted if it is radioactive and with a lifetime long enough to allow the storage of a significant sample that is extracted and counted. The approach is called *radiochemical technique* and has been pioneered by the *Brookhaven chlorine experiment* at Brookhaven, a tank with 4000 l of C_2Cl_4 (perchloroethylene). The Brookhaven experiment has been followed by the *Homestake chlorine experiment* with about 400,000 l of C_2Cl_4, at a depth of 4000 m.

The chlorine experiment is based on the reaction:

$$\nu_e + {}^{37}Cl \rightarrow {}^{37}Ar + e^- \tag{14.4}$$

that has a threshold at 814 keV. ${}^{37}_{18}Ar$ is chemically inert and has a half life of 35 days, thus it can be accumulated for long time intervals, two months in the original experiment. The extracted argon atoms decay through the inverse of the formation reaction into ${}^{37}_{17}Cl$ in an excited state, that disexcites emitting an X-ray with an energy of 2.82 keV. The chlorine experiment measured a discrepancy between the observed number of neutrinos and the theoretical predictions, the *solar neutrino problem*. The reaction is sensitive to ${}^8B$ and ${}^7Be$ neutrinos.

Another combination of elements suitable for radiochemical detection are gallium and germanium:

$$\nu_e + {}^{71}Ga \rightarrow {}^{71}Ge + e^- \tag{14.5}$$

The gallium reaction has a threshold at 233 keV and is sensitive to the pp neutrinos. The ${}^{71}_{32}Ge$ has an half life of 11.4 days. The *GALLEX* at Gran Sasso used 30 tons of a $GaCl_3$ solution and extracted germanium as $GeCl_4$ by bubbling nitrogen in the tank and removing the gas. The gallium experiment confirmed a deficit in the neutrino rate.

Large volume detectors are based on liquid scintillators or water Cherenkov detectors. The scintillators have no threshold and have a higher light yield, but the emission is almost isotropic and are worse from the point of view of the precision in the reconstruction of the direction [4]. The water Cherenkov detectors act as calorimeters, since the number of Cherenkov photons is proportional to the energy of the particle. The opening angle of the radiation allows a precise estimation of the particle tracks.

The *Super-Kamiokande* (SK) detector[6] is an underground (1000 m depth) water Cherenkov detector with a diameter of 39 m and an height of 42 m. The inner part of the tank is equipped with 11,200 photomultiplier tubes, while the outer part act as a shield against the atmospheric muons. The muons emit Cherenkov light while loosing energy in the water. The neutrino direction and energy are reconstructed by the morphology and intensity of the collected light. Electrons produce an electromagnetic shower, whose electrons and positrons emit Cherenkov light with a pattern more blurred than the muon one. The threshold of the experiment is about 5 MeV. The Super-Kamiokande detector observes solar neutrinos using the elastic scattering on electrons:

$$\nu + e^- \rightarrow \nu + e^- \tag{14.6}$$

The technique is sensitive to all neutrino flavors, even if the cross section for electron neutrinos is larger than those of the muon and tau neutrinos. The Super-Kamiokande detector had the ability to correlate the neutrino events with the Sun position and provided an independent confirmation of the solar neutrino deficit.

The *Sudbury Neutrino Observatory* (SNO)[7] provided the evidence of the ^{8}B neutrino oscillations. It was a spherical tank with a diameter of 12 m filled with 1000 tons of heavy water surrounded by a shield of 1500 tons of normal water. About 9500 photomultiplier tubes were used to measure the Cherenkov light of electrons produced in the neutrino reactions. The first step in detection was the elastic scattering on electrons, followed by the charged current reaction on the deuteron d with a threshold of 1.44 MeV:

$$\nu_e + d \rightarrow e^- + p + p \tag{14.7}$$

or a neutral current reaction with the emission of a 2.2 MeV photon:

$$\nu + d \rightarrow \nu + p + n \tag{14.8}$$

The neutrino oscillations were confirmed by the *Kamiokande Liquid scintillator ANtineutrino Detector* (KamLAND),[8] an instrument consisting of 1000 tons of liquid scintillator (with an high light yield) to investigate the positrons produced in the inverse β-decay by electron antineutrinos coming from nuclear reactors:

$$\bar{\nu}_e + p \rightarrow e^+ + n \tag{14.9}$$

The positron annihilation followed by the delayed emission of a 2.2 MeV photon emitted in the neutron capture are the signature of the electron antineutrino.

---

[6]http://www-sk.icrc.u-tokyo.ac.jp/index-e.html.

[7]http://www.sno.phy.queensu.ca/.

[8]http://www.awa.tohoku.ac.jp/kamland/.

The *Borexino* experiment[9] is based on a target made of 280 tons of liquid scintillator contained into a transparent vessel floating in a spherical tank of liquid, enclose in a tank of pure water as a shield and a veto against muons. The signal of neutrino events is detected by about 2000 photomultipliers.

14.4 Neutrino from Supernova Collapse

Neutrinos emitted in the supernova collapse have energies of the order of 10–20 MeV [1, 3, 5]. The largest contribution to the signal is given by electron neutrinos, that interact through the inverse β-decay. The energy of supernova neutrinos is larger than the threshold of the instruments for solar neutrino investigations, and the same detectors can be used. Water based detectors, such as Super-Kamiokande, and liquid scintillators systems, such as KamLAND, have a large number of protons for the inverse β-decay reaction.

Problems

14.1 Discuss the techniques used to detect low energy neutrinos.

14.2 Discuss the techniques used to detect high energy neutrinos.

References

1. Grupen, C. and Buvat, I: Handbook of Particle Detection and Imaging. Springer-Verlag Berlin Heidelberg (2012)
2. Katz, U. F., Spiering, Ch.: High-energy neutrino astrophysics: Statis and perspectives. Progr. Part. Nucl. Phys. **67**, 651 (2012)
3. Lèna, P. et al.: Observational Astrophysics. Springer-Verlag Berlin Heidelberg (2012)
4. Olive, K.A. et al. (Particle Data Group): Chin. Phys. C **38**, 090001 (2014)
5. Spurio, M.: Particles and Astrophysics - A Multi-Messenger Approach. Springer International Publishing, Switzerland (2015)

---

[9]http://borex.lngs.infn.it/.

Chapter 15
Gravitational Wave Astronomy

The direct detection of gravitational waves announced on 11 February 2016 has opened a new window onto the universe. The gravitational wave spectrum extends from some nHz to several kHz. This chapter will firstly discuss the acoustic detectors, the first historically developed. The laser interferometric detectors that have performed the discovery are described, together with the main noise sources: shot noise, thermal noise and seismic noise. Ground based interferometers are sensitive from a few Hz to some kHz, while space based interferometers are sensitive in the subHz region.

15.1 Orders of Magnitude

Gravitational waves [5, 7, 8, 11, 12] are ripples in spacetime that are produced by the accelerated motion of massive sources. The potential emitters are compact celestial objects: neutron stars, black holes, supernovae and binary systems in general. The direct detection of gravitational waves [1] was achieved with the observations of the event GW150914, the coalescence and merging of two black holes with masses of 36 and 29 $M_\odot$ that produced a final black hole with a mass of 62 $M_\odot$ and radiated 3 $M_\odot$.

The gravitational wave domain and the electromagnetic one are complementary astrophysical windows. Due to their weak interaction with matter, gravitational waves preserve the information of events that have a weak or missing electromagnetic counterpart, as shown by the first direct detection of the merging black holes. The weak interaction with the environment allows also to investigate the primordial Universe.

Gravitational waves produce a perturbation in the metric of spacetime that causes a tidal forces between two free test masses. The gravitational wave strength is described by the dimensionless amplitude, the *strain h*, that measures the fractional variation ΔL of the distance L separating the test masses:

© Springer International Publishing Switzerland 2017
R. Poggiani, *High Energy Astrophysical Techniques*,
UNITEXT for Physics, DOI 10.1007/978-3-319-44729-2_15

$$h = \frac{\Delta L}{L} \tag{15.1}$$

Gravitational waves have two possible polarizations, named *plus* (+) and *cross* (×). The gravitational wave strain is produced by the second derivative of the quadrupole moment Q of the source:

$$h \sim \frac{G}{c^4} \frac{\ddot{Q}}{r} \tag{15.2}$$

where r is the distance, G the Newton gravitational constant. The quadrupole moment can be approximated by $Q \sim M\delta^2$, where δ is the scale of deviation from symmetry. Thus $\ddot{Q} \sim Mv^2$, where v is the velocity. Assuming a binary system consisting of two neutron stars in the Virgo cluster (15 Mpc) with an orbital frequency of a few hundred Hz, the resulting strain will be $h \sim 10^{-21}$: this value has guided the building of large interferometers.

The spectrum of gravitational waves extends over several orders of magnitude, from 10^{-18} to 10^4 Hz:

- *Extremely Low Frequency*, 10^{-18}–10^{-15} Hz: primordial gravitational waves that can be observed in the polarization modes of the Cosmic Microwave Background (CMB)
- *Very Low Frequency*, 10^{-9}–10^{-3} Hz: gravitational waves from supermassive black hole binaries that can be detected in the fluctuations of the arrival times of the radio signal from pulsars
- *Low Frequency*, 10^{-5}–10^{-1} Hz: the potential sources are supermassive black hole binaries with masses in the range 10^4–10^9 $M_\odot$, Extreme Mass Ratio Inspirals (EMRI), binary stars, that can be detected by laser interferometers in space
- *High Frequency*, 1–10^4 Hz: the candidate sources are coalescing black holes and neutron star binaries, collapsing supernovae, pulsars; the region is investigated with ground based laser interferometers and with resonant detectors

The expected signal of some astrophysical sources of gravitational radiation is reported in Fig. 15.1

The coalescing binary systems are a potential source in different bands. In particular, black hole binaries with a wide range of masses are likely candidates. The ground based interferometer can investigate also the emission of gravitational waves from core collapse supernovae, that produces a burst wave, and the continuous emission of slightly asymmetric pulsars at twice the rotation frequency.

The output of a gravitational wave detector is the strain time series $h(t)$ described, in the frequency domain, by the *amplitude spectral density* $\tilde{h}(f) = \sqrt{S_h(f)}$, where $S_h(f)$ is the *power spectrum*, the square modulus of its Fourier Transform. The amplitude spectral density has the units of $\frac{1}{\sqrt{Hz}}$. An additional quantity of common use is the dimensionless *characteristic strain*, $h_c(f) = \sqrt{f S_h(f)} = \sqrt{f}\tilde{h}(f)$, the root mean square signal in a band width f around the frequency f. The noise sources are described by a noise power spectrum $S_n(f)$ or the noise density $h_n(f) = \sqrt{S_n(f)}$. All noise sources are assumed to be uncorrelated and are added in quadrature.

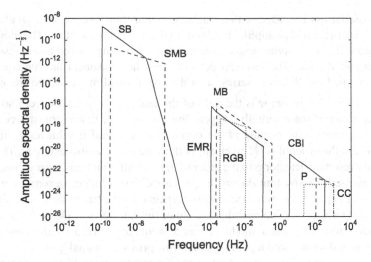

Fig. 15.1 The spectrum of gravitational wave sources; data from http://rhcole.com/apps/ GWplotter/. SB = Stochastic Background; SMB = Supermassive Binaries; EMRI = Extreme Mass Ratio Inspirals; MB = Massive binaries; RGB = Resolvable Galactic Binaries; CBI = Compact Binary Inspirals; P = Pulsars; CC = Core Collapse Supernovae

Gravitational wave detectors presently in operation belong to two main families: *resonant detectors*, mechanical systems whose resonance modes are excited the by incident waves; *interferometric detectors*, where the relative position of test masses is sensed by a laser and measured by the interference fringes.

15.2 Resonant Detectors

The first gravitational wave detectors have been the *resonant detectors* pioneered by Weber since the 1960s [5, 7, 8, 11, 12]. The resonant detectors are cylindrical bars with a high quality factor and a resonant frequency in the kHz region. The typical scales are a length of a few meters and a mass of some tons. Gravitational waves could excite the vibration modes of the bar, that are sensed by transducers glued on its surface. The simple model of a resonant detector is a pair of masses attached to a spring with a resonance frequency f_r. The excitation of the bar can be described by a damped harmonic oscillator with a frequency f_r and a quality factor Q. The resonant detectors are usually shaped as cylindrical bars and can oscillate at the lowest longitudinal mode, $f_r = \frac{v_s}{2L}$, where v_s is the sound velocity in the material and L the bar length. The efficiency of a bar is given by the ratio of the absorbed energy to the incident energy:

$$\chi \sim \frac{8GM}{\pi c} \frac{v_s^2}{c^2} \qquad (15.3)$$

The oscillation is sensed with a transducer that transforms it into an electrical signal. Then the signal is amplified and sent to the acquisition system. The transducer can be modeled by a spring-mass system, with a frequency tuned to the resonance frequency of the bar. The resonant detector and the transducer are the equivalent of two coupled oscillators in series, with the two normal modes at the frequencies $f_\pm = f_r(1 \pm \frac{w^{\frac{1}{4}}}{2})$, where w is the ratio of the masses of the transducer and of the bar. The incident wave initially triggers the bar motion, that is transferred to the transducer through the beating of the normal modes, until it achieves a vibration amplitude of the order of $w^{-\frac{1}{2}}$, much larger than the original displacement. The band width of a resonant detector is of the order of $f_r w^{\frac{1}{2}}$. Different transducer technologies have been used. Passive transducers rely on capacitive or inductive sensors, where a variation of capacity or inductance is proportional to the bar motion. The signal is amplified by a SQUID (Superconducting Quantum Interference Device). Parametric transducers use a resonator modulated at a frequency driven by the variation of capacity of inductance and a pump oscillator to provide a signal gain.

The resonance of bars can be excited by different sources of noise. The seismic noise from the ground vibrations is reduced by several orders of magnitude by multistage stack suspensions, with normal and internal modes far from the resonance frequency of the bar.

The spectral density of the *thermal noise* is:

$$S_{th} = k_B T \frac{\omega M}{Q} \tag{15.4}$$

The contribution of thermal noise is minimized by operating the bar at cryogenic temperatures and selecting materials with high quality factors, that should have at the same time a high speed sound to maximize the energy absorption. The traditional material for cryogenic bars is Al 5056 or, more rarely, niobium. The typical longitudinal length is some meters, corresponding to a resonance frequency in the kHz region, while the mass is of the order of tons. The minimum detectable strain, the ratio of the displacement to the bar length, is of the order of 10^{-16} for operation at room temperature and one order of magnitude smaller for operation at liquid helium temperature, 4.2 K, still larger than the expected astrophysical signals. The adopted solution is to secure a snapshot of the slowly damped oscillation after the excitation, that has a time scale $\tau_d = \frac{1}{2\pi f_r Q}$, using a short integration time τ_i. The spectral density of thermal noise is correspondingly rewritten:

$$S_{th} = 2k_B T \frac{\tau_i}{\tau_d} \tag{15.5}$$

The readout system introduces the *electronic noise*. The transducer can be modeled as a two port system described by a 2×2 impedance matrix Z. The inputs are force and velocity, while the outputs are current and voltage. The elements Z_{12} and Z_{21} represent the sensitivity of the transducer and the back action on the resonator, respectively. A transducer is a two way device and a fluctuating current could induce a fluctuating force on the bar. The back action noise is:

$$S_{back} = \frac{|Z_{12}|^2}{2M} I(\omega) t_{int} \tag{15.6}$$

where $I(\omega)$ is the current noise at the output of the transducer. The following stage is the amplifier, that introduces a noise related to the voltage noise $V(\omega)$ at the transducer exit:

$$S_{ampl} = \frac{2M}{|Z_{12}|^2} \frac{V(\omega)}{t_{int}} \tag{15.7}$$

Cryogenic resonators have a *standard quantum limit*:

$$h_{SQL} \sim \frac{1}{L} \sqrt{\frac{\hbar}{2\pi f_r M}} \tag{15.8}$$

The original Weber antenna operated at room temperature. The first cryogenically cooled bars joined into the network International Gravitational Event Collaboration (IGEC)[1]: Auriga (Legnaro, Italy), Allegro (Baton Rouge, USA), Explorer (CERN), NAUTILUS (Frascati, Italy), NIOBE (Perth, Australia). They first four antennas are made of Al 5056, while the last one is made of niobium. The IGEC network searched for event in coincidences, without finding. The cryogenic detectors achieved a minimum strain sensitivity of about 1×10^{-21} Hz$^{-\frac{1}{2}}$.

The technique of resonant detectors has been recently revamped, with the concept of spherical antenna. The sphere is isotropic and allows multiple independent channels for the analysis of the signals, to determine the direction of the gravitational wave. The target sensitivity is 1×10^{-22} to 1×10^{-23} Hz$^{-\frac{1}{2}}$. *MiniGRAIL*[2] is a cryogenic CuAl sphere with a diameter of 68 cm, a resonance frequency of 2.9 kHz and a bandwidth of about 200 Hz, that will operate at 20 mK.

15.3 Interferometric Detectors

The interferometric detectors are based on the configuration of the *Michelson interferometer* [5, 7, 8, 11, 12]. The layout of an interferometer with Fabry-Perot cavities in the arms and dual recycling (power and signal recycling) is shown in Fig. 15.2. The laser beam is split into two orthogonal beams by a beam splitter (BS). Each beam travels along a separate arm and is reflected by a mirror at the arm end. The beams return to the beam splitter and are sent to a photon detector. The relative change in the length of the two arms produced by the passage of a gravitational wave is measured by the variation of light intensity at the detector. The interferometers are broad band instrument, with a range of frequencies larger than that of resonant detectors, allowing the possibility of observe a large variety of astrophysical events.

---

[1] http://igec.lnl.infn.it/
[2] http://www.minigrail.nl/.

Fig. 15.2 Layout of a laser interferometer with Fabry-Perot cavities in the arms and dual (power and signal) recycling; *BS* is the beam splitter, *IX, EX* are the input and end mirrors of the first arm, *IX, EY* are the input and end mirrors of the second arm, *PR* is the mirror for power recycling, *SR* the mirror for signal recycling

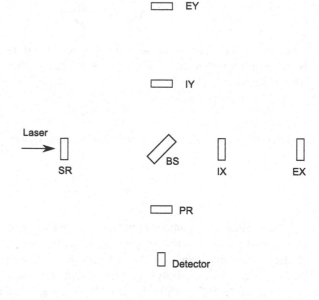

The variation ΔL of the length L is a phase shift:

$$\Delta\phi = 4\pi \frac{\Delta L}{\lambda} \tag{15.9}$$

The interferometers are operated in *dark fringe* and are stabilized by measuring the change in intensity at the output and feeding the signal to a transducer that changes the position of one of the mirrors. The variation in the length of the interferometer arms is extracted from the signal fed back to the transducer. The laser should be stabilized in power and in frequency. The optical quality of the mirrors must be very high, with losses of the order of some parts per million.

The strain measured by an interferometer is a combination of the two polarizations:

$$h(t) = f^+ h_+(t) + f^\times h_\times(t) \tag{15.10}$$

where f^+, $f^\times$ are the interferometer beam pattern responses to the two polarization states. The antenna pattern of a single interferometer is shown in Fig. 15.3. The response function is almost omnidirectional: interferometers are suitable for surveying the sky, but they cannot localize the position of an event as standard telescopes or radio telescopes. The position of the candidate source is estimated by combining the information of different instruments and the difference in the arrival times, as will be discussed in Chap. 17.

There are practical constraints to the physical length of the interferometer arms, set by the curvature of the Earth and by the cost of the infrastructures: we will see that the interferometer must operate in ultra high vacuum. Thus, the real baselines are 3–4 km. The interference signal is measured by a photon detector and is a measure of

Fig. 15.3 Antenna pattern
of a gravitational
interferometer, based on
https://github.com/tobin/
Peanut/blob/master/peanut.
m

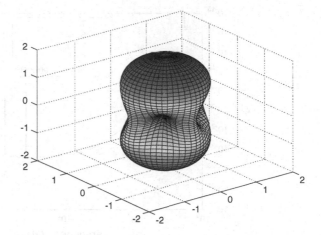

the relative change of the distance between test masses. If the optical elements (beam splitter, mirrors) are suspended as pendula, they will behave as free test masses above the pendulum frequency. We will see in the following that the pendulum concept is fundamental to tackle other aspects of the detection. The variation of the arm length is proportional to the interferometer arm length. Assuming a strain $h \sim 10^{-21}$ and an arm length of L of 4 km, the expected variation is $\Delta L \sim 4 \times 10^{-18}$ m, one thousandth of the size of a nucleus.

An interferometer shows several different noise sources, related to fundamental physics processes, such as thermal noise or shot noise, or to environmental conditions, such ad the seismic noise, or to technical aspects, such as the laser fluctuations in power and in frequency. The strain produced by residual gas with density ρ made of molecules with polarizability α and mean velocity $\bar{v}$ is [12]:

$$h_{rg} = 2^{\frac{5}{2}} \pi^{\frac{5}{4}} \frac{\alpha}{\lambda^{\frac{1}{4}} L^{\frac{3}{4}}} \sqrt{\frac{\rho}{\bar{v}}} \exp\left(-\frac{\sqrt{2\pi\lambda L} f}{\bar{v}}\right) \tag{15.11}$$

The achievement of an high sensitivity requires the operation of the whole interferometer in ultra high vacuum, at a pressure of 10^{-9} mbar. The laser beams must travel inside evacuated pipes with kilometric lengths and large diameters, of the order of one meter. The ultra high vacuum should have a low partial pressure of hydrocarbons to avoid the contamination of the optical elements. The cost of the vacuum system is a large part of the total cost of the interferometer.

The Virgo sensitivity curve is shown in Fig. 15.4. The contribution of the main noise sources is explicitly shown and will be discussed in the following, together with the technical solutions.

The optical path traveled by the beams can be increased by having them travel back and forth several times in the arms, i.e. by folding the optical path. There is an intrinsic limitation to the effective arm length. When it approaches the the wavelength

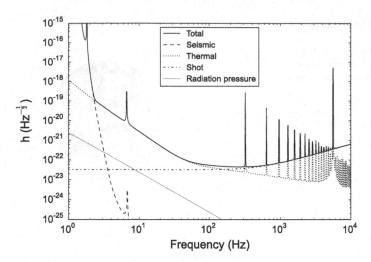

Fig. 15.4 Main noise sources in the Virgo interferometer; data from http://www.virgo-gw.eu/senscurve/

of the gravitational wave, the gravitational effect sensed after the many round trips will be different from the one at the instrument input, with a drop in sensitivity. The first solution to optical folding is the Herriot delay line, that uses two concave mirrors with a suitable radius of curvature to have the incident light exiting after N round trips, that correspond to an optical path $2NL$. The laser beam position moves at each trip along an annulus with radius Nw_0, where w_0 is the beam diameter. The total number of trips is limited by the reflectivity of the mirrors, that must have large size to accommodate the beam reflections. The second solution for folding is the use of Fabry-Perot cavities in the arms of the interferometer. The Fabry-Perot cavity is made of two plane mirrors with high reflectivities at a distance L. The light reflected from the cavity is composed of the radiation directly reflected by the entrance mirror and by the radiation coming back from the cavity interior. The reflection coefficient of the Fabry-Perot cavity depends on the mirror reflectivities r_1, r_2 and on the transmittivity t_1 of the input mirror:

$$r_{FP} = -r_1 + \frac{r_2 t_1^2}{1 - r_1 r_2 e^{-2ikL}} \tag{15.12}$$

when the condition $kL = n\pi$, with n integer, the cavity is in *resonance* and the laser power is trapped as in an optical delay line. Any change in the cavity length induces a phase shift close to resonance. The optical performances of a Fabry-Perot is described by the sharpness at the resonance, the finesse $\mathcal{F} = \frac{\pi \sqrt{r_1 r_2}}{1 - r_1 r_2}$. The equivalent number of round trips is $N = \frac{2\mathcal{F}}{\pi}$, with a typical value of the order of a few hundreds, that gives a total optical path of some hundreds km. The laser beam in a Fabry-Perot cavity is bouncing back and forth at the same position on the mirror, thus mirrors can be smaller than those used for the optical delay lines, but the optical properties must be very high since the radiation must traverse them at the beginning and at the end of the folded path. The operation at resonance of the Fabry-Perot cavities sets strong requirements on the stability of the cavity length and on the laser frequency. The input and end mirrors for the cavities in the two arms of the interferometer in the layout of Fig. 15.2 are labeled IX, EX and IY, EY, respectively.

Laser is a source of different noises. The *shot noise* is due to the fluctuations of the number of photons detected at the recombination port. The amplitude spectral density of the shot noise in an interferometer with optical folding is [12]:

$$h_{sn}(f) = \frac{1}{NL}\sqrt{\frac{\hbar c \lambda}{2\pi P}} \tag{15.13}$$

where N is the number of round trips, P is the laser power. The use of high power lasers reduces the contribution of the shot noise. The interferometers in operation have adopted the Nd:YAG lasers, with a wavelength of 1.064 μm. There is an upper limit to the usable power, since the radiation pressure produces a mechanical motion of the mirrors. The spectral density of the *radiation pressure noise* is [12]:

$$h_{rp}(f) = \frac{N}{MLf^2}\sqrt{\frac{2\hbar P}{\pi^3 c \lambda}} \tag{15.14}$$

where M is the mirror mass. This noise is minimized by using large test masses. The combination of the two noises by addition in quadrature produces the *optical readout noise*; radiation pressure noise dominates at low frequencies, while shot noise at high frequencies. The readout noise at fixed frequency is minimum when the two contributions are equal, i.e. for a power $P = \frac{\pi c \lambda M f^2}{2N^2}$. The optimal power sets the *Standard Quantum Limit*, that depends only on the Planck constant and on the basic parameters of the interferometer, the mass of the mirrors and the arm length [12]:

$$h_{SQL} \sim \frac{1}{\pi f L}\sqrt{\frac{\hbar}{M}} \tag{15.15}$$

The standard quantum limit is much smaller than the sensitivity of the first and second generation detectors.

The effective power of the laser can be increased to reduce the shot noise, using the *power recycling* technique [12]. The operation of the interferometer at the dark fringe uses only a small fraction of the laser light, that is sent back to the input port. A partially transmitting mirror (PR) inserted in front of the laser, before the beam splitter, reflects the unused light inside the interferometer, increasing the usable power by about two orders of magnitude. In addition to the power recycling technique described above, the sensitivity of an interferometer can be improved in a narrow band with the *signal recycling* technique [12]. A partially reflecting mirror (SR) with a suitable reflectivity placed at the output of the interferometer sends part of the radiation back for interference and transforms the instrument into a resonant cavity.

The interferometers in operation use the combination of the power recycling and signal recycling, the *dual recycling*.

In addition to the shot noise and the radiation pressure noise, there are displacement noises, that constrain the physical scale of the instrument. Several noise sources produce a displacement x of the test masses, that is equivalent to a strain $h \sim \frac{x}{L}$. The seismic noise from ground vibrations produces a motion of the optical elements of the interferometer, simulating a genuine gravitational signal. The thermal noise excites the vibrations of the interferometer elements. The seismic noise is the dominant noise at low frequency, while thermal noise sets the limit at intermediate frequencies.

The *seismic noise* is produced by the ground motion excited by different processes, that range from earthquakes to wind. The noise injected in the interferometer makes the optical elements vibrate. The displacement spectrum of the seismic noise above a few Hz is [12]:

$$x_{seism} = A_s \left(\frac{10 \, Hz}{f} \right)^2 \frac{cm}{\sqrt{Hz}} \qquad (15.16)$$

where A_s is in the range 10^{-7} to 10^{-6}. The noise is larger than the gravitational signals by several orders of magnitude. The effect of the seismic motions on the test masses is reduced using active or passive suspension systems as isolators. A mass M connected to a suspension point through an elastic constant k has a transfer function proportional to $\frac{f_0^2}{f_0^2 - f^2}$, where $f_0 = \frac{1}{2\pi}\sqrt{\frac{k}{m}}$. Above the characteristic frequency the isolation is proportional to $\left(\frac{f_0^2}{f^2} \right)$. The interferometer mirrors are suspended as pendula to behave as the free test mass, with a characteristic frequency below the band width of the interferometer. A single pendulum does not provide sufficient isolation in the horizontal direction. The test mass requires additional isolator stages, a cascade of n oscillators providing an attenuation of $\left(\frac{f_0^2}{f^2} \right)^n$. The suspension systems must provide also attenuation along the vertical direction, because of the Earth curvature and unavoidable asymmetries in the apparatus.

The *gravity gradient noise* [12] is a Newtonian effect produced by the fluctuations of the density of the ground at the observatory site, triggered by atmospheric or seismic activity. The isolator systems described above are not effective, since this noise is due to the direct coupling of the test mass to the seismic noise motions of the ground close to the interferometer. The coupling is equivalent to an elastic constant with a resonance at the frequency of the order of $\sqrt{G\rho}$, where ρ is the density of the ground.

The dominant noise at intermediate frequencies is the *thermal noise* produced by the vibration modes of the suspension systems, of the wires supporting the mirrors and to the normal modes of the test masses. Each mode can be described by a damped oscillator at temperature T. The contribution of viscous damping by the medium where masses are moving are minimized by having the interferometers operating in ultra high vacuum. The residual damping is produced by the anelasticity of the suspension material. The elastic constant k is replaced by a complex constant $k(1 + i\phi(\omega))$, where $\phi(\omega)$ is the *loss angle* that depends on the frequency, but very weakly for most materials. The loss angle is of the order of 10^{-3} for steel, 10^{-7} for fused silica and of 10^{-9} for silicon or sapphire. The physical interpretation of the loss angle is the reciprocal of the quality factor Q of the damped oscillation. The power spectral density of the thermal noise of an oscillator with mass M and elastic constant k is given by the Fluctuation-Dissipation Theorem [12]:

$$x_{th}^2 = \frac{4 k_B T k \phi}{\omega[(k - M\omega^2)^2 + k^2 \phi^2]} \qquad (15.17)$$

Thermal noise shows a peak at the resonance frequency of the oscillator. If the dissipation does not depend on frequency, the power spectral density behaves as ω^{-1} below resonance and as ω^{-5} above. Thermal noise can be minimized by operating the system at low temperatures and/or using large masses and/or minimizing the loss factor. The last approach has been thoroughly pursued in the interferometers in operation, by optimizing the materials of the suspensions and of the test masses.

The test masses are suspended to the last stage of the suspensions by one or two wire loops or by crystal fibers. The first solution is cheaper, requiring only clamps on the upper part, but has the problem of friction between the wires and the mirror. The second solution requires the bonding of the fiber ends to the last stage of the suspension and to the mirror.

The suspension thermal noise is mainly produced by the pendulum mode of the test mass, with a typical frequency of about 1 Hz, below the sensitivity band of the interferometer. Mounting the test mass as a pendulum offers another advantage, in addition to providing the free test mass condition. The intrinsic loss factor is weighted by the ratio of the elastic and gravitational forces, thus the effective loss factor for the pendulum mode is [12]:

$$\phi_p = \phi \frac{n_w}{2Mgl} \sqrt{I T_w Y} \qquad (15.18)$$

where n_w is the number of wires, l the pendulum length, I, T_w are the wire cross section moment of inertia and the wire tension, Y the Young modulus of the material. The gain factor over the intrinsic ϕ factor is one or two orders of magnitude.

The violin string modes are produced by the wires supporting the test mass. The frequencies of the violin modes are given by:

$$f_n = \frac{n}{2l} \sqrt{\frac{T_w}{\rho_l}} \qquad (15.19)$$

where ρ_l is the linear mass density of the wires. The violin modes are an harmonic series starting at some hundreds Hz that shares the same loss factor of the pendulum mode and produces the very narrow peaks in the sensitivity curve.

The normal modes of the test masses are an additional source of noise. The first normal mode of the test mass is at a frequency of about $f_1 = \frac{v_s}{2d}$, where v_s is the sound velocity in the material and d the mirror diameter. The thermal noise of all internal modes at frequencies f_n with intrinsic losses ϕ_n [12]:

$$x^2_{mirror} = \frac{8k_B T}{2\pi f} \sum_n \frac{\phi_n}{4\pi^2 M f_n^2} \qquad (15.20)$$

The mirrors are designed to have normal modes with frequencies lying above the band width of the interferometer, i.e. at several kHz. The standard material is fused silica, with loss factors in the range 10^{-6} to 10^{-8}.

The quality factors of the vibration modes of the suspension and of the test mass must be optimized, since high values reduce thermal noise, but increase the injection of the seismic noise. Thus the upper stages of suspensions are built to have normal modes with low quality factors, while the lower stages require high quality factors [12].

The interferometers have a large number of control systems, both at the local and global level. Local controls deal with the damping of the normal modes of the suspension system. Global controls keep the interferometer locked and the cavities in resonance. The main control is a feedback loop that manages the differential arm length, keeping the interferometer in *dark fringe*. The differential error signal is used to send a control signal to the mirrors to correct the effect of any motion, including that produced by a gravitational wave [12].

Laser interferometry in space investigates the gravitational waves in the frequency range from 10^{-3} to 10^{-5} Hz, not accessible on Earth because of the seismic noise. It is an extension of the terrestrial techniques to much larger arm lengths, of the order of millions kilometers, by the use of Doppler *spacecraft tracking*, to detect gravitational waves in the deciHertz region and below. The original concept of the *LISA* interferometer consisted of three spacecraft at the vertexes of an equilateral triangle with a side of five million kilometers, in an heliocentric orbit lagging 20° behind the Earth. Each spacecraft will contain a pair of test masses, acting as a shield against external disturbances and following their motion. The distance of the spacecrafts will be measured using laser ranging. Each spacecraft receiving the laser beam will reproduce a reflected beam by transmitting a new laser beam locked in phase to the incident beam. The phase of the returning beam will be compared with the phase of the main laser to extract the redshift of the back and forth trip. The frequency noise of the laser will be canceled using the technique of *time delay interferometry*, a combination of the six signals in the back and forth trips in the three arms of the interferometer. The dominant noise contribution will be shot noise at high frequency and acceleration noise of the test masses at low frequencies.

15.4 Pulsar Timing

The technique of *Pulsar timing* investigates the region from nanohertz to mHz [7]. The pair of test masses are the Earth and a distant pulsar. Pulsars are very precise clocks, with an accuracy of the same order of atomic clocks. Gravitational waves affect the arrival time of the pulses from the pulsars. The times measured on Earth are converted to the Solar System barycenter, with a correction that depends on the relative position of the Earth and the pulsar. The difference between the observed pulse time and the predicted pulse time is named *timing residual*. Old pulsars are better candidates than young pulsars, whose timing residuals are often large. The observation times are of the order of months.

15.5 Gravitational Observatories

The first generation of ground based interferometric detectors has shown the technical feasibility of large instruments for gravitational wave detection and have set upper limits to several classes of sources. The experiments have often operated in coincidence. The collaborations have agreements for the operation as a network, the sharing of data and the joint analysis. The first generation instruments have arm length from 300 m to 4 km and often operated in coincidence:

- *Laser Interferometer Gravitational-wave Observatory* (LIGO)[3]: two separate interferometers with an arm length of 4 Km located at Hanford, WA and Livingstone, LA, in the USA, and an additional 2 Km arm length interferometer colocated at Hanford
- *Virgo*[4]: an interferometer with an arm length of 3 km, located in Cascina, Pisa, Italy
- *GEO 600*[5] (GEO 600): interferometer with an arm length of 600 m, close to Hannover, Germany
- *TAMA*[6]: interferometer with an arm length of 300 m in Japan

LIGO and Virgo have used Fabry-Perot cavities and power recycling, while GEO 600 and TAMA have chosen the strategy of Michelson interferometry with dual recycling. In the initial LIGO interferometers the test masses were suspended to a passive isolator consisting of alternating layers of steel and Viton. The initial Virgo interferometer adopted an active strategy, the superattenuator, a cascade of pendula for horizontal attenuation equipped with metal cantilever blades and magnetic antisprings for vertical attenuation. The initial mirror suspension were based on steel wires ($\phi \sim 10^{-3}$), and have been progressively replaced by fused silica monolithic fibers ($\phi \sim 10^{-7}$).

---

[3]https://www.ligo.caltech.edu/.
[4]https://www.virgo-gw.eu/.
[5]http://www.geo600.org/.
[6]http://tamago.mtk.nao.ac.jp/spacetime/tama300_e.html.

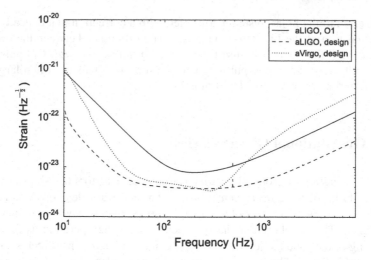

Fig. 15.5 Sensitivity of Advanced LIGO during O1 run [9] and final (data from [10]) and sensitivity of Advanced Virgo [3]

The second generation interferometers [4–8, 11, 12], Advanced LIGO[7] and Advanced Virgo,[8] have been designed to improve the sensitivity by different techniques: mirrors with better reflectivities and larger masses to reduce both shot and thermal noise; the use of the signal recycling; improved reduction of seismic noise; higher power lasers. The design sensitivity of Advanced LIGO and Advanced Virgo and the sensitivity of LIGO interferometers during the observing run O1 are shown in Fig. 15.5.

The Advanced LIGO interferometers have performed the first observing run O1 form September 2015 to January 2016, achieving the first direct detection of gravitational wave. The Advanced LIGO interferometers use 40 kg test masses suspended to the final stage of a quadruple pendulum system supported by an active isolator. The test masses are made of fused silica and are suspended with fused silica fibers. Advanced Virgo should be operative before the end of 2016. It will use the super-attenuators, very low loss mirrors and an improved vacuum level in the tube. The interferometer *KAGRA*,[9] with an arm length of 3 km, is expected to start operation in 2018. The interferometer will be built in an underground mine and will use sapphire mirrors cooled at 20 K to reduce the thermal noise.

---

[7]https://www.ligo.caltech.edu/.

[8]http://public.virgo-gw.eu/language/en/.

[9]http://gwcenter.icrr.u-tokyo.ac.jp/en/.

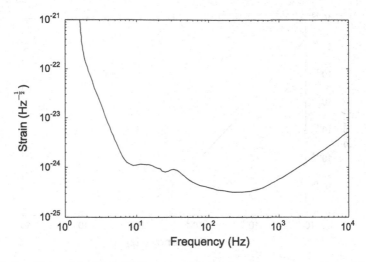

Fig. 15.6 Sensitivity of the Einstein Telescope; data from http://www.et-gw.eu/etsensitivities

The third generation ground based interferometers, such as *Einstein Telescope*[10] will allow an improvement in sensitivity by one order of magnitude. The future interferometers will be underground to reduce the effects of the seismic noise. The operation of the suspension systems and of the mirrors at cryogenic temperatures will mitigate the thermal noise; the lower temperature reduces also the loss factor of several materials. The primary solution to improve shot noise is the increase of the laser power. The technique of *light squeezing* is an alternative technique. The squeezing of the phase fluctuations increases the amplitude fluctuations, thus shot noise can be reduced at the expense of the radiation pressure noise, with a better sensitivity at high frequencies. The sensitivity of Einstein Telescope is shown in Fig. 15.6.

The eLISA is a proposed mission for space based interferometry, scheduled for launch in 2034. The mission will use three spacecraft (one mother and two daughters) that form a Michelson interferometer configuration. The spacecrafts will follow independent heliocentric orbits forming an equilateral triangle following the Earth at a distance in the range 10°–30°. Gravitational sensor units will be installed on the spacecrafts, one for each daughter and two for the mother. The units will contain a free falling test mass. A telescope will transmit light along the interferometer arm and receive the signal from the other ones. The interference of the received light with the light of a reference laser is a measure of the Doppler shift from the relative motion of the spacecrafts, that is summed to the local measurement of the displacement between the test mass and the spacecraft. The strain sensitivity of eLISA is shown in Fig. 15.7. The sensitivity is dominated by the residual acceleration noise

---

[10]http://www.et-gw.eu/.

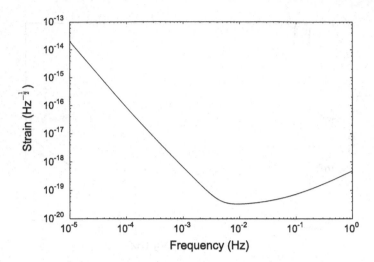

Fig. 15.7 Sensitivity of the space based interferometer eLISA; data from [2]

of test masses at low frequencies, by the measurement noise (including shot noise) at intermediate frequencies and by the arm length response at high frequencies.

Problems

15.1 Discuss the main sources of noise in a ground based interferometer.

15.2 Discuss the possible solutions for the reduction of the seismic noise in interferometers.

15.2 Discuss the possible solutions for the reduction of the thermal noise in interferometers.

References

1. Abbott, B. P. et al.: Observation of Gravitational Waves from a Binary Black Hole Merger. PRL **116**, 061102 (2016)
2. Amaro-Seoane, P.: Low-frequency gravitational-wave science with eLISA/NGO. Class. Quantum Grav. **29**, 124016 (2012)
3. Acernese, F. et al.: Advanced Virgo: a second-generation interferometric gravitational wave detector. Class. Quantum Grav. **32**, 024001 (2015)
4. Bassan, M.: Advanced Interferometers and the Search for Gravitational Waves. Springer Heidelberg New York Dordrecht London (2014)
5. Blair, D.: The Detection of Gravitational Waves. Cambridge University Press (1991)
6. Blair, D. G., Howell, E. J., Ju, L., Zhao, C.: Advanced Gravitational Wave Detectors. Camdridge University Press (2012)

7. Creighton, J. D. E., Anderson, W. G.: Gravitational-Wave Physics and Astronomy - An Introduction to Theory, Experiment and Data Analysis. WILEY-VCH Verlag GmbH & Co. KGaA, Germany (2011)
8. Hawking, S. W., Israel, W.: Three Hundreds Years of Gravitation. Cambridge University Press (1987)
9. https://dcc.ligo.org/LIGO-T1200307/public
10. https://dcc.ligo.org/LIGO-T0900288/public
11. Maggiore, M.: Gravitational Waves, Volume I, Theory and Experiments. Oxford University Press (2008)
12. Saulson, P. R.: Fundamentals of Interferometric Gravitational Wave Detectors. World Scientific Publishing Co. Prc Limited (1994)

Chapter 16
The Dark Side of the Universe

The recent cosmological investigations have shown that dark matter and dark energy have a major role in the composition of the Universe. The existence of dark matter has been inferred by the anomalous rotation curves in spiral galaxies and by the gravitational potential of galaxy clusters. Several candidates for dark matter have been proposed. This chapter presents the techniques for the direct detection of dark matter particles (Weakly Interacting Massive Particles, WIMP) through the scattering on nuclei. Then the indirect techniques to detect dark matter based on the emission of gamma rays in the annihilation of WIMPs and the positron excess in cosmic ray observations are presented. A concise discussion of dark energy and the related investigations will be given.

16.1 Dark Matter

The *dark matter* [1–3, 5] has been originally proposed by Zwicky in 1933 to explain the velocity dispersion of the galaxies in the Coma cluster. There are several astrophysical evidences for the existence of dark matter [2, 5]. The rotation curves of spiral galaxies suggest the presence of a dark halo extending beyond the range of visible matter. The contribution of the dark matter should be larger in dwarf spheroidal galaxies [2]. Observations of galaxy clusters suggest the presence of a larger amount of matter than that belonging to the stellar component. The cluster galaxies are inside a gas emitting X-rays, since it is warmed by the combined gravitational potential of the galaxies, the gas and dark matter. The galaxy clusters are relevant also for the gravitational lensing, that bends the light emitted from objects behind them. Additional constraint are set by the angular power spectrum of the temperature anisotropies of the Cosmic Microwave Background, that depend on several cosmological parameters, among them the cosmological constant Λ, the matter density Ω_m, the baryon density Ω_b, the Hubble constant H_0. The combination of the

© Springer International Publishing Switzerland 2017
R. Poggiani, *High Energy Astrophysical Techniques*,
UNITEXT for Physics, DOI 10.1007/978-3-319-44729-2_16

different cosmological measurements [3] suggests that the Universe is flat and is composed of 4 % of baryons, 20 % of non baryonic dark matter and 76 % of dark energy (see Sect. 16.2).

The strongest dark matter candidates are the *Weakly Interacting Massive Particles* (WIMP), predicted by several models of new physics [2, 5]. In supersymmetric (SUSY) models, the WIMP is the *neutralino*, with a predicted mass in the range from 10 to 10^3 GeV.

The WIMPs can annihilate or scatter on standard matter [5]. The *annihilation* process produces normal matter particles through the coupling of the WIMPs to nuclei. The neutralino annihilates into a fermion-antifermion pair, whose spins are oppositely directed. The helicity factor of the reaction is proportional to the squared mass of the fermion, thus the neutralino preferentially decays into heavy quarks or leptons. The annihilation cross section is of the order of 1 pb [5]. The final products can be searched with the *direct detection experiments*. The *elastic scattering* of WIMPs onto a nucleus produces a recoil of the latter, whose energy can be measured. The cross section for elastic scattering is smaller than that of annihilation by some orders of magnitude [5].

The searches for dark matter are direct or indirect. The direct searches rely on the physical detection of dark matter particles crossing the detection system. Indirect searches rely on the detection of the products of the annihilation.

The estimated local energy density of dark matter is about 0.3–0.4 GeV cm^{-3}, while the velocity distribution is assumed to be Maxwellian, with an average velocity of 220–230 km s^{-1} [2, 5]. The local density of the WIMPS, assuming a WIMP mass of about 100 GeV, is a few thousands per cubic meter. The Earth motion inside the dark matter distribution is the composition of its orbital motion and of the Sun motion in the Galaxy, that gives the name *WIMP wind* to the observed flux at the Earth. The Earth orbital motion produces the seasonal modulation in the collected flux, up to 7 %, achieving a maximum on June, 2nd.

16.1.1   Direct Detection of Dark Matter

The *direct detection of dark matter* [1–3, 5] relies on the elastic scattering of WIMPs on nuclei and has the typical signature of the seasonal modulation of the signal produced by the Earth motion around the Sun. The average kinetic energy of the WIMPs is of the order of tens keV [2], larger than the binding energy of nuclei in solids. The direct detection methods measure the energy of the recoiling nucleus. The interaction can occur through a spin dependent or a spin independent process. In the former case, only unpaired nucleons participate to the interaction, while in the latter there is a coherent contribution from all nucleons. The interaction rate for unit mass is proportional to A for spin dependent and to A^3 for spin independent reactions, where A is the atomic mass of the material [2]. The direct searches for dark matter use targets with high atomic masses. The energy spectrum of the nuclear recoil has a peak in the low energy region, below tens keV, assuming a target with $A \sim 100$

and WIMPs with a mass of tens GeV. The expected event rates are smaller than 1 events/(day kg) and smaller than 10^{-3} events/(day kg) for the spin independent and spin dependent couplings [2, 5], demanding for large mass detectors.

The seasonal modulation is a tool to discriminate between WIMP produced events and background events. A part of background should show a seasonal modulation described by a cosine function with a period of 1 year and with a maximum amplitude and epoch of maximum as discussed above. The sensitivity of experiments for direct dark matter detection depends on the detector mass and on the observation time [2], thus large mass detectors and long exposure times are needed. The ideal WIMP detector should have a low threshold for nuclear recoils and a low intrinsic background. The contribution of the environmental radioactivity to the background is minimized by operating the dark matter detectors at underground sites and using shielding against charged and neutral particles.

The WIMPs can be detected without discriminating the nuclear recoils, but looking for the annual modulation of the signal. The technique has been pioneered by the *DAMA/LIBRA*[1] experiments at the Gran Sasso Laboratories, that achieved a statistics of 290 and 530 kg · year. The two experiments consisted of arrays of NaI(Tl) crystals, with a total mass of about 87 and 233 kg, respectively. The experiments have observed the annual modulation, but with a candidate neutralino unfitting the theoretical prediction.

The solid state detectors operating at cryogenic temperatures have thresholds lower than 10 keV and intrinsic energy resolution capability. The *CoGeNT* experiment[2] uses the ionization signal of high purity germanium. The low threshold is useful to investigate the background at very low energies, that can be produced by light WIMPs.

Another approach to WIMP detection relies on the combination of different detection processes to identify the nuclear recoils. The mechanisms for particle detection are ionization, scintillation and phonon (heat) production. Using two of them, it is possible to identify the nuclear recoil events [2], that are more efficient in producing phonons than in producing an ionization or a scintillation signal. The pairing of detection techniques has produced different classes of hybrid detectors. Instruments based on the detection of light and heat are arrays of semiconductor diodes with the addition of a phonon sensor. The diodes are large size Si or Ge detectors, while the heat sensors are doped semiconductors or thin superconducting films. The detector assemblies operate at temperatures of the order of tens of milliKelvin. The *EDELWEISS*[3] experiment uses ionization heat bolometers, that measure the temperature variation caused by the energy released in the interaction and the number of produced electrons and achieve a high rejection of the background. The *CDMS*[4] experiment at the Soudan Underground Laboratory (USA) uses low temperature cylindrical Si and Ge crystals equipped with electrodes for the ionization measurement and supercon-

---

[1] http://people.roma2.infn.it/~dama/web/home.html.

[2] http://cogent.pnnl.gov/.

[3] http://edelweiss.in2p3.fr/.

[4] http://cdms.berkeley.edu/.

ducting transition edge sensors on one base for the detection of phonons with high spatial resolution. The time difference between the ionization and photon signals allow the identification of nuclear recoils. The *scintillating bolometers* are arrays of scintillating materials equipped with a phonon sensor and operating at very low temperatures. The light signal is converted into an heat signal using additional bolometers with dedicated phonon sensors. The *CRESST*[5] experiment uses cryogenic detector modules based on the scintillating $CaWO_4$ crystal, with a sapphire wafer for measuring the light mounted on one side and a tungsten superconducting thermometer to measure the temperature increase. The family of instruments based on the detection of the light and of the charge use liquefied noble gases as absorbers, that are effective both as ionization and scintillation detectors. The ratio of the signals in charge and in scintillation allows to separate the electrons and nuclear recoils. The nuclear and electron recoils are discriminated from their different light yield.

The experiment *XENON10*[6] is based on UV sensitive photomultiplier tubes to measure the light signals of events in the Xenon target with the capability of three dimensional reconstruction.

The *Darkside50* experiment at LNGS[7] is a liquid argon Time Projection Chamber inside a vessel of liquid scintillator, to reduce the background.

16.1.2   *Indirect Searches of Dark Matter*

The *indirect dark matter detection* [1–3, 5] explores the mechanism of annihilation of WIMPs and the related products. The dark matter density is predicted to rise at the Galactic center, making it a natural target for searches. Additional targets are possible dark matter clumps in the galactic halo and massive celestial objects that trap the particles. Among all products of annihilation, the neutral particles are suitable probes, since their spectrum is not affected by the magnetic fields. The annihilation into gamma rays preserves the information about the direction of the source, in the Galaxy or extragalactic. The WIMP annihilation in the halo can produce a continuum spectrum of gamma rays and monoenergetic photons. The small branching ratios of final states containing one or more photons prevents searches of lines in the spectra. The main background is given by the gamma rays produced by the standard interactions of cosmic rays with the interstellar medium, through the intermediate step of neutral pion production.

The annihilation of WIMPS can produce antiparticles that must be disentangled from the antiparticles produced in the interaction of the cosmic rays with the interstellar medium. Positrons are a minority component of the cosmic rays. The PAMELA telescope (Chap. 13) has observed an excess of the positron flux in the region from 10 to 200 GeV, that could be associated to the annihilation of dark matter.

---

[5]http://www.cresst.de/darkmatter.php.

[6]http://xenon.astro.columbia.edu/XENON10_Experiment/.

[7]http://darkside.lngs.infn.it/ds-50/.

16.2 Dark Energy

The current cosmological observations suggest that a large fraction of the energy density of the Universe is made of *dark energy* [2, 3] responsible for the accelerated expansion of the Universe. Two supernova surveys have independently found evidence of the accelerated cosmic expansion. Several theoretical possibilities have been considered to explain the accelerated expansion [2]: a cosmological constant Λ, a scalar field, such as the quintessence or the k-essence, or modifications of General relativity.

The dark energy leaves an imprint in several astrophysical observables [1–3, 5]. The observations of type Ia supernovae rely on their luminosity distance, that is estimated from the measured apparent magnitude and the absolute magnitude considering them as standard candles. The presence of dark energy produces a larger luminosity distance than the one in a flat Universe without dark energy. The luminosity distances of observed supernovae was larger than the values predicted by a model of flat Universe without dark energy, suggesting $\Omega_m \sim 0.3$, $\Omega_\Lambda \sim 0.7$. The presence of the dark energy has an impact on the anisotropies in the temperature of the Cosmic Microwave Background. The position of the peaks in the spectrum of anisotropies is determined by the evolution of the expansion. The dark energy produces a shift in the position of the peaks. The data of the WMAP experiment have confirmed the energy density of dark energy found in supernovae observations. The Baryon Acoustic Oscillations (BAO) are produced by the pressure waves propagating in the photon-baryon fluid that leave a signature in baryon perturbations. The technique uses the photometric redshifts produced by photometric surveys and the observations of spectroscopic surveys. The measurement of the Baryon Acoustic Oscillations provides absolute distance and is the complement to the observations of the type I supernovae, that provide ratio of luminosity distances. The weak gravitational lensing of galaxies at high redshifts by clustered matter provides a snapshot of the cosmic structure growth, as the Redshift Space Distortions (RSD). The direct measurement of the Hubble constant H_0 constraints the value of the critical density.

The most part of dark energy experiments uses imaging and spectroscopic observations in the optical or the near infrared. The details of the imaging and spectroscopic techniques can be found in [1, 4]. A group of ground based experiments is based an the systematic photometric surveying of the sky. The *Dark Energy Survey* (DES)[8] aims to perform high precision measurements to be used for searches with weak lensing, baryonic acoustic oscillations and cluster based techniques. A byproduct of the survey is the detection of type Ia supernovae whose distance can be combined with the other measurements. The *Large Synoptic Survey Telescope* (LSST)[9] will image the Southern sky, providing systematic measurements of clusters, weak lensing and photometric BAO targets. The ground based spectroscopic surveys investigate the redshift-space distribution of galaxies. The *Baryon Oscillation Spectroscopic Sur-*

---

[8]https://www.darkenergysurvey.org/.
[9]https://www.lsst.org/.

vey (BOSS)[10] uses fiber fed spectrographs to reconstruct the space distribution of targets for estimation of the baryon acoustic oscillations and redshift space distortions. The *Hobby-Eberly Telescope Dark Energy Experiment* (HETDEX)[11] targets galaxies with Lyman-α emission.

The future observatories will be based in space. The *WFIRST*[12] observatory will perform imaging surveys for studying clusters and weak lensing and spectroscopic observations to emission line galaxies for baryon oscillation and redshift distortion studies.

Problems

16.1 Discuss the direct methods for dark matter search.

16.2 Discuss the indirect methods for dark matter search.

References

1. Lèna, P. et al.: Observational Astrophysics. Springer-Verlag Berlin Heidelberg (2012)
2. Matarrese, S., Colpi, M., Gorini, V., Moschella, U.: Dark Matter and Dark Energy. Springer Dordrecht Heidelberg London New York (2011)
3. Olive, K.A. et al. (Particle Data Group): Chin. Phys. C **38**, 090001 (2014)
4. Poggiani, R.: Optical, Infrared and Radio Astronomy - From Techniques to Observation. Springer (2016), Book DOI 10.1007/978-3-319-44732-2
5. Spurio, M.: Particles and Astrophysics - A Multi-Messenger Approach. Springer International Publishing, Switzerland (2015)

---

[10]http://cosmology.lbl.gov/BOSS/.

[11]http://hetdex.org/.

[12]http://wfirst.gsfc.nasa.gov/.

Chapter 17
Observing in High Energy Astrophysics

This chapter discusses the preparation of observations in high energy astrophysics, the signal to noise ratio for different telescope systems, and the techniques of data analysis for the different astronomies previously presented.

17.1 Observation Planning

The preparation of observations in high energy astrophysics is helped by the tools available at the site of the experiment or mission. Simulator tools are available to estimate the count rates and the exposure times. An example is the *Portable, Interactive Multi-Mission Simulator* (PIMMS)[1] for X-ray astronomy. In the following, we will discuss the signal to noise ratio in different instruments as a guide to estimate the orders of magnitude involved.

17.1.1 Signal to Noise Ratio in Gamma Ray Astronomy Telescopes

The sensitivity of a gamma ray telescope, i.e. the minimal detectable flux, has been discussed by [11, 13]. The minimum detectable flux at an energy E is given by [11, 13]:

$$\Phi_{min}(E) = n \frac{\sqrt{S(E) + B(E)}}{A_e T} = \frac{n}{2A_e(E)T} \left(n + \sqrt{n^2 + 4B(E)} \right) \qquad (17.1)$$

---

[1] https://heasarc.gsfc.nasa.gov/docs/software/tools/pimms.html.

© Springer International Publishing Switzerland 2017
R. Poggiani, *High Energy Astrophysical Techniques*,
UNITEXT for Physics, DOI 10.1007/978-3-319-44729-2_17

where n is the number of standard deviations of the fluctuations of the background (for example 3), $S(E)$, $B(E)$ are the counts from the source and the background, $A_e(E)$ the effective area, T the observation time. The counts from the source are a function of the minimum flux, $S(E) = \Phi_{min} A_e T$. Usually the background contribution is dominant ($n \ll 2\sqrt{B(E)}$), thus the minimum detectable flux is:

$$\Phi_{min}(E) = n \frac{\sqrt{B(E)}}{A_e(E)T} \tag{17.2}$$

The counts from the background depend on its energy spectrum, $\frac{d\phi_b}{dE}$ [11, 13]:

$$B(E) = \frac{d\phi_b}{dE} A_e(E) \Delta E \Delta \Omega T \tag{17.3}$$

where ΔE is the energy range of the observation or the energy resolution and $\Delta \Omega$ is the element of angular resolution.

17.1.2   Signal to Noise Ratio in Atmospheric Cherenkov Detectors

The ground based atmospheric Cherenkov detectors have a signal to noise ratio determined by the flux Φ_c of the Cherenkov photons collected by the photomultiplier system [12]:

$$S = \Phi_c A \eta \tag{17.4}$$

where A is the collection area, η the quantum efficiency of the photon detector. The Cherenkov pulse involves some photons per square meter in a time interval of a few nanoseconds.

The background from the night sky is:

$$B = \Phi_{nb} A \eta \Omega T \tag{17.5}$$

where Φ_{nb} is the night sky background, Ω the field of view of the instrument, T the observation time. The signal to noise ratio is:

$$SNR = \frac{S}{\sqrt{B}} = \frac{\Phi_c A \eta}{\sqrt{\Phi_{nb} A \eta \Omega T}} \tag{17.6}$$

The minimum detectable signal is inversely proportional to the signal to noise ratio, thus the energy threshold of the instrument is:

$$E_{threshold} \propto 1/\Phi_c \sqrt{BT\Omega/\eta A} \tag{17.7}$$

The interplay between the area and the solid angle allows to balance smaller areas with the reduction of the solid angle, lowering the threshold of the instrument.

17.1.3 Signal to Noise Ratio in X-Ray Telescopes

An example of the signal to noise ratio in a coded mask instrument has been presented by [7]. Generally the observations are dominated by the instrumental background. The detection sensitivity is:

$$S = \frac{2n\sqrt{B(E)T}}{\varepsilon(E)(1 - \tau_0)\tau_1 AT\Delta E} \tag{17.8}$$

where n is the number of standard deviations, $B(E)$ the background, T the observation time, ΔE the energy band, ε the product of the detection and imaging efficiencies, τ_0 the mask transparency, τ_1 the hole transparency.

17.2 Observing and Data Analysis Techniques

The data of high energy astrophysics are called *events* and contain at least the position, the arrival time and the pulse height or energy of the photon. Additional information depends on the specific mission, but often include the characteristics of the detector and the satellites, the flags for the quality of the event etc. In detectors with intrinsic energy resolution, such as the CCDs, the energy is defined by the charge stored in pixels, called *pulse height amplitude* (PHA). The time intervals corresponding to the storage of the events are called *Good Time Intervals* (GTI).

17.2.1 Gamma Ray Data Analysis

The analysis of data from gamma ray telescopes starts from the typical observation procedure [8]. The gamma ray telescope is pointed in the direction of the target for a time interval t_{on}, recording N_{on} photons. The telescope is then pointed for a background measurement for a time interval t_{off}, counting N_{off} photons. The number of photons of the background included in the counts during the source observation are:

$$N_B = \alpha N_{off} \tag{17.9}$$

where $\alpha = \frac{t_{on}}{t_{off}}$ is the ratio of the times spent in observing the source and the background. The number of photons from the source are:

$$N_S = N_{on} - N_B \tag{17.10}$$

with a variance:

$$\sigma(N_s) = \sqrt{N_{on} + \alpha^2 N_{off}} \tag{17.11}$$

The significance of the observation is given by the Poisson statistics:

$$S = \frac{N_s}{\sigma(N_s)} = \frac{N_{on} - \alpha N_{off}}{\sqrt{N_{on} + \alpha^2 N_{off}}} \tag{17.12}$$

By using all data during t_{on}, t_{off} to estimate the background, the significance becomes:

$$S = \frac{N_{on} - \alpha N_{off}}{\alpha(N_{on} + N_{off})} \tag{17.13}$$

The significance can be estimated by the maximum likelihood method [8], assuming as null hypothesis that there are no extra sources and that all detected photons correspond to the background. The significance of the observation becomes:

$$S = \sqrt{2}\left[N_{on} ln\left(\frac{1+\alpha}{\alpha}\left(\frac{N_{on}}{N_{on} + N_{off}}\right)\right) + N_{off} ln\left((1+\alpha)\left(\frac{N_{off}}{N_{on} + N_{off}}\right)\right)\right]^{\frac{1}{2}} \tag{17.14}$$

The Eq. 17.14 can be used also with a low number of observed counts.

17.2.2 Atmospheric Cherenkov Data Analysis

The images secured by Cherenkov telescopes contain the Cherenkov radiation of interest and the background from of the night sky [5]. Usually the science target is monitored for a fixed interval, during the ON scan, then a reference region of the sky is monitored during the OFF scan, looking for differences in the number of gamma ray candidates. The pedestal (signal at zero illumination) and the related rms of the single channels are estimated by injecting artificial events in the detection system when it is not recording the Cherenkov light of the targets. The pedestal level is subtracted from the images. The pedestal variance is an indicator of the night sly background and is a tool to find the positions of the objects. The gains of the pixels are estimated by illuminating the camera system with a uniform and pulsed source of radiation and used to apply the flat field correction to the camera data. The reduced images are later cleaned by setting to zero the pixels governed by the sky noise background. Pixels whose content exceeds a threshold that is a few times the sky brightness are kept for further analysis. The neighboring pixels are analyzed again and accepted if they exceed a second, less restrictive threshold. The final image is the combination of the first group of pixels and of the neighboring pixels.

The gamma/hadron separation relies on the shape of the shower image [5], an ellipse (Chap. 12). The moments of the image are estimated using the information of the amplitudes in the pixels. The zero-th moment is the sum of all signals that have been cleaned according to the steps described above. The first and second order moments describe the position and the shape of the image. The main parameters of the shower image are the length, the width (major and minor semi-axis) and the orientation parameters.

17.2.3 X-Ray Data Analysis

The X-ray detectors record the signatures of individual photons [2]. The raw data are lists of events, where each event is described at least by the position, the arrival time and the pulse height or energy. The scientific data are obtained through dedicated pipelines specific to each mission. Some operations are common to all analysis procedures. The first step is the reconstruction of the position of the event on the sky, starting from the position on the detector in the focal plane of the telescope and the direction of pointing. The second step is the determination of the genuine nature of the event. The neighboring pixels within a 3×3 or 5×5 matrix around the event are analyzed and compared to the preset threshold, retaining those above and assigning a quality flag, the grade. The following step is the determination of the event energy. The procedure is different, according to the instrument. In detectors with intrinsic energy resolution, such as the CCDs, the energy is defined by the charge stored in pixels is called *pulse height amplitude* (PHA). The energy of an event is defined by the sum of the charges in the pixels with a content above the threshold (PHAS). The observed counts $O(x, y, pi)$ in the pixel with coordinates x, y and energy in the bin pi are given by:

$$O(x, y, pi) = \int \int \int \int R(x, y, pi, x_p, y_p, E, t) I(x_p, y_p, E, t) dx_p dy_p dEdt$$

(17.15)

where R is the *instrumental response function*, $I(x_p, y_p, E, t)$ the flux of the source at the sky position x_p, y_p at the energy E and the time t.

In the *imaging* the image is built by summing over the pi bins; the instrumental response is defined as:

$$R_{image}(x, y, x_p, y_p, E, t) = \sum_{pi} R(x, y, pi, x_p, y_p, E, t)$$

(17.16)

The response for images is decomposed as a function of the *Point Spread Function* (PSF) of the observing system, the probability distribution of event positions from a point source, and of the *Exposure Map* (EA):

$$R_{image}(x, y, x_p, y_p, E, t) = PSF(r, \theta, x_p, y_p, E) EA(x_p, y_p, E, t)$$

(17.17)

where $r = (x - x_p)^2 + (y - y_p)^2$, $\theta = \arctan \frac{y - y_p}{x - x_p}$. The Full Width Half Maximum of the Point Spread Function describes the spatial resolution of the instrument, typically in the range of some arc sec. The PSF depends on the energy. The exposure map is the effective area of the telescope at the position (x_p, y_p) at the energy E and at the time t.

The spectra of sources almost constant in time and over the investigated spatial region are built by the binning over a region according to:

$$O(pi) = \int \left(\int \int \int \int R(x, y, pi, x_p, y_p, E, t) dx_p dy_p dx dy dt \right) I(E) dE$$

(17.18)

The response is factorized into the *Ancillary Response Function* (ARF) and the *Redistribution Matrix File* (RMF). The ARF describes the effective area of the telescope as a function of energy. The RMF describes the probability that a photon of a certian energy will be detected in a chosen energy channel.

The *HEASoft*[2] suite is the combination of several tools for the analysis of X-ray data, among them *XRONOS* for timing analysis (light curve creation, period search, power spectra estimation), *XIMAGE* for image display and analysis (image processing, source detection, PSF generation, extraction of light curves and spectra), *XSPEC* for spectral fitting of data to a large number of physical models.

17.2.4   Gravitational Wave Data Analysis

An interferometer for gravitational wave detection is similar to a vibration sensor, producing a time series measuring the interference of the laser beams that have traveled along the arms [3, 4, 6, 9, 10]. The data analysis of gravitational wave interferometers shows remarkable differences compared to other astronomical observations. Gravitational detectors have an almost isotropic response, being able to detect events over almost the whole sky: they act as all sky survey instruments. A network of detectors is needed to reconstruct the event position with good angular resolution. The interferometers in the network operate as a single instrument that monitors the sky. A candidate event corresponds to a signal in two detectors at least. The arrival time at each interferometer gives the information about the arrival direction.

The signal to noise ratio is defined starting from the output $w(t)$ of the interferometer [10]. A signal in the time series can be identified by comparing it with a template $s(t)$ through the cross-correlation function:

$$w \star s(t) = \int w(\tau) s(t + \tau) d\tau$$

(17.19)

The cross correlation will be maximum when the template coincides with the signal. If the noise in the detector has a Gaussian probability distribution, the

---

[2]https://heasarc.gsfc.nasa.gov/lheasoft/.

cross-correlation will also have a Gaussian distribution. A candidate signal will have a large value of the correlation, with a signal strength defined by:

$$S^2 = |w \star s(t)| \tag{17.20}$$

The noise is:

$$N^2 = \sqrt{< (w \star s(\tau))^2 >} \tag{17.21}$$

The *signal to noise ratio* is defined as [10]:

$$SNR = \sqrt{S^2/N^2} \tag{17.22}$$

In the general case of non white noise, the inverse Fourier Transform of the *matched filter* with the transfer function:

$$G(f) \propto \frac{e^{-2\pi i f t_0} S^*(f)}{N^2(f)} \tag{17.23}$$

provides the template [10], where the time parameter is related to the interval of the time series considered for the cross correlation. An estimation of the signal to noise ratio for a signal with an amplitude s_0 is given by [10]:

$$SNR \sim \frac{s_0}{\sqrt{n^2(f)\Delta(f)}} \tag{17.24}$$

where the noise has been estimated using the product of the average noise power spectrum and the band width of the observation. The output of the interferometer $w(t)$ should be compared with the rms value of the noise. A candidate signal should be selected if it is above a chosen threshold, high enough to have a small probability of being exceeded by pure noise [10]. The coincident detection of the signal in more than one detector is a tool against spurious effects. If the probability of crossing the threshold at instrument i is p_i, the probability of simultaneous crossing of the threshold is the product $\prod p_i$. The instruments of the network should be at large distances to avoid common environmental disturbances. The time delay between interferometers separated by a distance D is:

$$\Delta t = \frac{D}{c} \tag{17.25}$$

The delay is of the order of 7 ms for the two LIGO detectors and of about 20 ms between LIGO and Virgo. The coincident observations with different instruments is requested to determine the position of the source [10]. With two detectors at distance D it is possible to reconstruct the event within an annulus, whose axis is the prolongation of the baseline joining the detectors. The equivalent of the declination coordinate for the annulus position is $\frac{c\Delta t_{12}}{D}$, where Δt_{12} is the difference in the arrival time at the two detectors. Three detectors define two independent annuli, that restrict the region of the event to a patch in the sky.

The search for periodic gravitational wave is performed by estimating the Fourier Transform of the interferometer output and looking for peaks in the power spectrum [10]. The same pattern should be observed by all instruments of the network. Periodic gravitational waves have a clear signature, the amplitude and frequency modulation caused by the relative motion of the Earth and the source.

The stochastic background can be detected by cross correlating the signal of two separated interferometers, since it should be more correlated than their intrinsic noises [10]. The distance between the two detectors cannot be arbitrarily large, since cross correlation is not negligible for waves with a wavelength larger than the distance. On the other hand, detectors too close or sharing the same vacuum system show correlated noises.

The analysis of GW150914, the first detected event [1], is available as a tutorial at https://losc.ligo.org/s/events/GW150914/GW150914_tutorial.html. The main steps are reported here as an example of gravitational wave data analysis. The strain data presented in Fig. 17.1 show a dominant low frequency noise.

Fig. 17.1 Strain time series of the Hanford (H1) and Livingston (L1) interferometers; the zero time is the event time

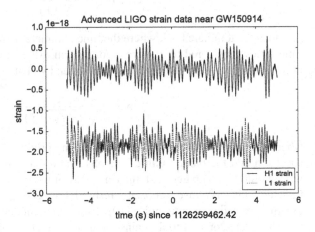

Fig. 17.2 Strain time series of the Hanford (H1) and Livingston (L1) interferometers after whitening and band passing, and numerical template generated with the estimated parameters of the binary system; the data of L1 have been shifted by 7 ms and inverted (see [1] for details)

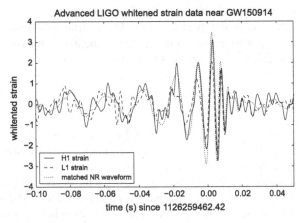

The signals can be extracted by whitening the data, i.e. by dividing them in the frequency space by the noise amplitude spectrum to deal with the noise at low frequency, and by band passing them with a Butterworth filter to cut the high frequency noise. Both signals show the signature expected for the coalescence of binary black holes: an oscillation with increasing frequency corresponding to the inspiral phase followed by the merger and ringdown phases. The signal is consistent with the prediction of general relativity, as shown by the numerical relativity template waveform for the signal expected from a pair of black holes with masses of about 36 and 29

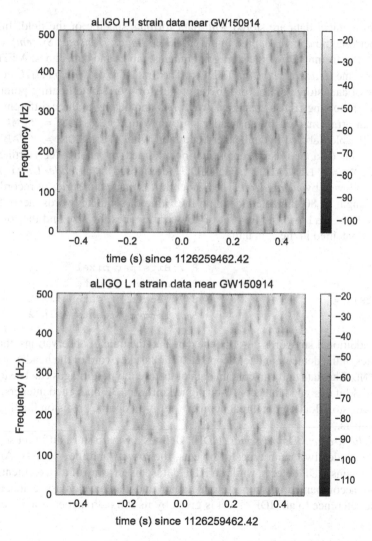

Fig. 17.3 Spectrograms of the strain time series of the Hanford (H1) and Livingston (L1) interferometers after whitening

solar masses, that merge into black hole of 62 solar masses, at a distance of about 410 Mpc (Fig. 17.2).

The time-frequency spectrogram of the whitened data show the typical *chirp* signature of the coalescence of two compact objects (Fig. 17.3), with an increase in the frequency of the signal in time.

17.3 Software

The astronomical data are stored with data formats specific of the field. Imaging and spectroscopic data are stored in the *(Flexible Image Transport System)* (FITS) format.[3] The format allows the storage of images, tables and data cubes. A FITS file contains some *Header/Data Units* (HDUs). The first HDU or *Primary HDU* contains an array of data stored as 1, 2 or 4 byte integers or 4 or 8 byte floating point numbers. The following HDUs are called *extensions* and belong to three different types: image, an array marked by an header starting with XTENSION = 'IMAGE '; an ASCII table, labeled by an header beginning with XTENSION = 'TABLE '; a table with data in binary representation, defined by an header starting with XTENSION = 'BINTABLE'. Each HDU is composed of an *Header Unit* in ASCII format, followed by a *Data Unit*. The header unit is made of a series of records with a fixed length of 80 characters defining the value of some keywords, according to KEYNAME = value. The first set of keyword defines the size and the format of the data contained in the data unit:

```
BITPIX  =                    8 / Bits per pixel
NAXIS   =                    2 / number of data axes
NAXIS1  =                  768 / length of data axis 1
NAXIS2  =                  768 / length of data axis 2
```

The additional keywords describe the date and the time of observations, the characteristics of the detector and telescope etc. The last keyword of the header is necessarily END. The data unit can contain data of different types. Images can be stored as arrays of 8-bit unsigned integers 16-bit signed integers, 32-bit signed integers, 32-bit single precision floating point real numbers or 64-bit double precision floating point real numbers.

The *Hierarchical Data Format* (HDF)[4] defines a way to store different scientific data. It includes two main types of objects, the *groups* and the *datasets*. An HDF group contains HDF objects, while an HDF dataset is an array of data elements; both types are accompanied by supporting metadata and may have an associate attribute list. The reference to an HDF object is given by its full path name, as a directory in UNIX.

---

[3]http://fits.gsfc.nasa.gov/fits_primer.html.
[4]https://www.hdfgroup.org/HDF5/doc/H5.format.html.

SAOImage DS9 (DS9)[5] is an application for the visualization of FITS images and tables, with the possibility of defining regions of interest, using multiple frames, loading images from external catalog.

The NASA Heasarc site provides different software tools to view and manipulate FITS files. *Fv*[6] is a software to view and interactively edit FITS files. The elements of an image or of a table can accessed with tools similar to those of spreadsheets, building the graph of one variable against the other or building histograms of variables. *FTOOLS*[7] and *CFITSIO* are suites of programs for reading, writing and editing FITS files.

Problems

17.1 Discuss the signal to noise ratio in the observation of high energy photons.

17.2 Discuss the techniques of data analysis in X-ray astronomy.

17.3 Discuss the techniques of data analysis in gamma ray astronomy.

17.4 Discuss the techniques of data analysis in gravitational wave astronomy.

References

1. Abbott, B. P. et al.: Observation of Gravitational Waves from a Binary Black Hole Merger. PRL **116**, 061102 (2016)
2. Arnaud, K. A., Smith, R. K., Siemiginowska, A.: Handbook of X-ray Astronomy. Cambridge University Press (2011)
3. Blair, D.: The Detection of Gravitational Waves. Cambridge University Press (1991)
4. Creighton, J. D. E., Anderson, W. G.: Gravitational-Wave Physics and Astronomy - An Introduction to Theory, Experiment and Data Analysis. WILEY-VCHVerlag GmbH & Co. KGaA, Germany (2011)
5. Fegan. D. J.: γ-hadron separation at TeV energies. J. Phys. G: Nucl. Part. Phys. **23**, 1013 (1997)
6. Jaranowski, P., Kròlak, A.: Analysis of Gravitational-Wave Data. Cambridge University Press (2009)
7. Lèna, P. et al.: Observational Astrophysics. Springer-Verlag Berlin Heidelberg (2012)
8. Li, T.-P., Ma, Y.-Q.: Analysis methods for results in gamma-ray astronomy. Ap. J. **272**, 317 (1983)
9. Maggiore, M.: Gravitational Waves, Volume I, Theory and Experiments. Oxford University Press (2008)
10. Saulson, P. R.: Fundamentals of Interferometric Gravitational Wave Detectors. World Scientific Publishing Co. Prc Limited (1994)
11. Schoönfelder, V.: The Universe in Gamma Rays. Springer-Verlag Berlin Heidelberg, (2001)
12. Weekes, T.: Very High Energy Gamma-Ray Astronomy. Institute of Physics Publishing, Bristol and Philadelphia (2003)
13. Zhang, S. N., Ramsden, D.: Statistical data analysis for gamma-ray astronomy. ExA **1**. 145 (1990)

---

[5]http://ds9.si.edu/site/Home.html.

[6]http://heasarc.gsfc.nasa.gov/docs/software/ftools/fv/.

[7]http://heasarc.gsfc.nasa.gov/docs/software/ftools/ftools_menu.html.

Chapter 18
Conclusions

High energy astrophysics is a part of the multiwavelength and multimessenger astrophysics. This chapter discusses the impact of high energy observational astrophysics in the building of a picture of Universe at all scales.

The possibility of observing in different electromagnetic bands and with different information carrier is a major advantage for the investigation of the Universe. High energy astrophysical phenomena are often associated to compact objects and to transient events and can involve high energy photons, neutrinos, cosmic rays, gravitational waves and also low energy photons. Astrophysical observations of an object are the combination of investigations requiring different technologies and different analysis methods, often in coincidence.

The event GW150914, that provided the direct evidence of gravitational waves [1], is an example of multimessenger astronomy, since it has involved electromagnetic and neutrino follow-ups. Preliminary estimates of the time and the location in the sky of the event have been shared with a network of ground and space based observatories covering the whole electromagnetic spectrum: radio, optical, near-infrared, X-ray and gamma-rays [2]. Since the event was a binary black hole merger, no electromagnetic counterpart is expected. The observations of the neutrino follow-up [3] has found no candidates that could be associated to the gravitational wave event.

The future astrophysical observations with photons, cosmic rays, neutrinos, gravitational waves will continue to integrate the information of different information carriers to tackle the unsolved problems of astrophysics.

© Springer International Publishing Switzerland 2017
R. Poggiani, *High Energy Astrophysical Techniques*,
UNITEXT for Physics, DOI 10.1007/978-3-319-44729-2_18

References

1. Abbott, B. P. et al.: Observation of Gravitational Waves from a Binary Black Hole Merger. PRL **116**, 061102 (2016)
2. Abbott, B. P. et al.: Localization and Broadband Follow-up of the Gravitational-wave Transient GW150914. ApJ Letters **826**, L13 (2016)
3. Adrian-Martinez, S. et al.: High-energy neutrino follow-up search of gravitational wave event GW150914 with ANTARES and IceCube. Phys. Rev. D **93**, 122010 (2016)

Index

© Springer International Publishing Switzerland 2017
R. Poggiani, *High Energy Astrophysical Techniques*,
UNITEXT for Physics, DOI 10.1007/978-3-319-44729-2

Printed in the United States
By Bookmasters